$4 —
Ⓜ
6/23

D0038707

THE GRIZZLY BEAR

LIBRARY OF
JIM EATON

From a photograph, copyright, 1909, by J. D. Kerfoot

AT BAY

THE GRIZZLY BEAR

THE NARRATIVE OF A HUNTER-NATURALIST

BY
WILLIAM H. WRIGHT

FOREWORD BY Frank C. Craighead, Jr.

UNIVERSITY OF NEBRASKA PRESS
LINCOLN AND LONDON

THIS BOOK IS
DEDICATED
WITH THE RESPECT, ADMIRATION, AND AFFECTION
OF THE AUTHOR
TO THE NOBLEST WILD ANIMAL OF NORTH AMERICA,
THE GRIZZLY BEAR

Publishers on the Plains

UNP

Copyright, 1909, by Charles Scribner's Sons
Foreword Copyright © 1977 by the University of Nebraska Press
All rights reserved

First Bison Book printing: 1977
Most recent printing indicated by first digit below:
2   3   4   5   6   7   8   9   10

Library of Congress Cataloging in Publication Data
Wright, William Henry, 1856–1934.
   The grizzly bear.
   Reprint of the 1913 issue of the ed. published by
Scribner, New York.
   Includes index.
   1. Bear hunting.   2. Grizzly bear.   I. Title.
SK295.W74 1977   639'.11'74446   77–1772
ISBN 0–8032–0927–4
ISBN 0–8032–5865–8 pbk.

   This edition is reproduced from the 1913 edition published
by Charles Scribner's Sons, New York.

   Manufactured in the United States of America

FOREWORD

THE GRIZZLY BEAR, by William H. Wright, first published in 1909, is one of the best all-around books ever written on the grizzly bear. It is both highly informative and entertaining. Wright, an unusually keen observer with an inquiring mind, spent twenty-five years more or less constantly observing the grizzly. His interpretations of what he learned are cautiously and meticulously arrived at and are accurate even when compared with the findings of recent studies having the advantages of the modern tools and disciplines of science.

Wright began as a bear-hunter, and an extraordinarily successful one. He pitted his own strength, endurance, ingenuity, skill, knowledge, and craftiness against that of the grizzlies. He met them on nearly equal terms, killing most of his grizzlies with a one-shot Winchester and following them for days and weeks at a time, seeking them out, not drawing them in with baits. Only an exceptionally skilled hunter could have accomplished what he did in the course of a hunting career. His most remarkable achievement as a hunter was killing five grizzlies with five shots, which he called "the greatest bag of grizzlies that I have ever made single-handed." From this time on Wright's interest turned from killing grizzlies to studying them. He said, "If one really wishes to study an animal, let him go without a gun"; he set aside his rifle, took up photography,* and continued to learn all he could about the grizzly. His book shows a hunter becoming a naturalist: Wright first studied the grizzly in order to hunt him, then he came to hunt him in order to study him.

THE GRIZZLY BEAR is most valuable for the wealth of information it offers. It treats the early history of the grizzly as recorded by the white man and the life and escapades of James Capen ("Grizzly") Adams, and most important it recounts the true-life experiences of Wright himself. Although I have spent some eighteen years studying the grizzly, eight of them intensively, there are few points on which I would take issue with the accuracy of Wright's observations or his interpretations of what he saw. His scientific classification of grizzlies, in Chapter V, is obsolete. Wright's experience led him to believe that the grizzly's winter den is usually a natural cave; in some areas, however, nearly all dens are excavated and prepared by the bear. Wright believed that cubs would follow their mothers the second summer before being cast aside to shift for themselves. We have found that often the yearlings or "cubs" are weaned in spring; they may den with their mothers over a second winter, but the family will break up before the second summer.

In his opening autobiographical chapter the author states, "This is an animal much talked about but little studied. It is now well on its way toward extinction."

---

*Although by present-day standards the quality of the author's photographs leaves a good deal to be desired, they have been retained in the present edition for their historic interest.

The writing of this book was Wright's attempt to help prevent the inevitable decline of the grizzly. In his second chapter on the early history of the bear, he emphasizes this aim: "When my grandfather was born, the grizzly had never been heard of. If my grandson ever sees one it will likely be in the bear pit of a zoological garden." Wright's hunting and killing of grizzlies in an era when this was nearly universally accepted resulted in a book whose objective was to preserve the bear through better understanding by the general public. It is my hope that reprinting this book for wide distribution will continue to serve the cause to which Wright devoted his later life.

FRANK C. CRAIGHEAD, JR.
Moose, Wyoming

CONTENTS

PART I

HISTORICAL

PART II

MY EXPERIENCES AND ADVENTURES

Contents

PART III

CHARACTER AND HABITS OF THE GRIZZLY

ILLUSTRATIONS

Illustrations

PART I
HISTORICAL

I

AUTOBIOGRAPHICAL

THE object of this book is to place on record the facts I have gleaned, and the deductions I have drawn, from some twenty-five years of more or less constant observation of the grizzly bear. This is an animal much talked about, but little studied. It is now well on its way toward extinction. Our acquaintance with its life history is broken by many gaps and supplemented by many conjectures. Some of these gaps I believe myself able to fill; some of these conjectures I propose to examine and discuss. Before, therefore, venturing even tentatively to take the chair as a witness, I find it only right that I should, in so far as I am entitled to do so, qualify as an expert.

I was born in 1856 on a small farm in southern New Hampshire. This farm, like its neighbors, was little more than piles of rock fenced in by other piles of rock; and from the time I was twelve years old or so I worked on these piles of rock during the summer, and at coopering or lumbering during the winter, and in the saw-mills in the spring.

I had little schooling, and, indeed, little chance of it; but one of my earliest recollections is nevertheless con-

nected with a book.   I have often seen in the newspapers
and magazines replies of various persons of note to the
question, "What book has exerted the greatest influence
on your life?"   Most of these answers I notice are rather
hazy, but if I had ever been asked to reply to this question,
I should have been able to answer without any hesitation.
And my answer would have been, "The Adventures of
James Capen Adams, Grizzly Bear Hunter of California."
This book, by what chance I am unable to guess, was in
the town library of the village nearest our home.   I suppose that my father, thinking that its illustrations would
amuse his sons, had brought it home for us.   At any rate,
although I have no recollection of seeing it for the first
time, I remember my father's reading aloud from it in the
evenings, and our repeated request that he would get it
for us again.   I think that before we could read ourselves,
my brother and myself must have all but worn out the old
book looking at its pictures.

Along about this time, in the early sixties, when I was
some six or seven years of age, Barnum's Circus made a
tour of the New England States, and their posters exhibited the picture of a huge grizzly, which was advertised as
having been caught by this man Adams.   You can, perhaps, imagine the effect of this upon us youngsters, and
the condition we were in when the circus came to Nashua,
and our father consented to take us to see it.   That bear
is about the only thing I remember about my first circus.
I know I went back every few minutes to look at him; and
I can see him now, pacing backward and forward in his
great cage.   I can see just how his toe-nails looked and
can remember the exact color of him.   There is nothing

strange in the fact that I had already, boylike, dreamed of emulating Adams. It is even less strange that that evening I openly declared my resolution. The only strange thing about it is that I never changed my mind.

In those days, and in that country, fathers had an undisputed proprietary interest in their children until they came definitely of age. It never occurred to either side to question it, and while, as I grew toward manhood, I found myself more and more seriously determined to go west and become a hunter, I worked steadily on the home farm and at home pursuits till I was twenty-one. Then, since I had no means of my own, I went to Fitchburg, and got work in a machine-shop. I had already, at home, learned blacksmithing; and now, as I had had little opportunity of education, I went to night school. I worked, first and last, all over New England. I was soon getting the highest wages. I was never laid off when there was any work to be done, and whenever a shop closed down at one place, I got a letter of recommendation and went to the next town.

Meanwhile I had not lost sight of my real purpose. I had long owned a tent and a rifle. I had, whenever I could, gone into the woods. I knew well how to fend for myself in the open, and was something of a crack shot both at a mark and at game.

In 1883 (I was then working in Providence, R. I., for Brown & Sharp), a man came around the country hiring mechanics for a large machine-shop in Melbourne, Australia. The men, I remember, were to pay their own way out, and if they stayed eighteen months, were to be reimbursed. It was a far cry from the opportunity I had been looking for, but it interested me enough to send me to the ticket office to make inquiries. A man named Church was ticket agent at the time, and in the course of our conversation he told me that there were a number of people thinking of going west, and that if I would give him three weeks in which to advertise a trip, and would take charge of two car-loads of emigrants, he would give me a ticket to Portland, Oregon, and return.

Naturally I jumped at the chance, though I wished many a time before I was rid of my charges that I had paid my fare. One car-load of settlers was switched off at Chicago bound for San Francisco; the others, with myself in charge, went north-west to St. Paul and beyond. And I dropped the last of them at Mussel Shell, Montana, where there was a large sheep-raising industry. There were now four of us, a carpenter from Maine, a jeweller and a blacksmith from Providence, and myself. I was keeping Melbourne up my sleeve, but had determined to stop off in the West if I could find a locality where I would be apt to get a grizzly, and I had made inquiries all along the road from every one who was supposed to know the country. Of course I had heard the usual number and the usual kind of bear stories, but I was so repeatedly informed that in the hills surrounding Spokane, there were plenty of grizzlies, if any one had the nerve to hunt them, that in the end three of us, the jeweller, the blacksmith, and myself, got off there.

Spokane was then a town of about fifteen hundred inhabitants. I had brought a camp outfit with me from the East and, when we got off the train, I secured a wheelbarrow, wheeled my trunk and my other belongings down

to the river bank about half a mile out of town, set up my
tent, and went hunting—for a job. I had twelve dollars
and fifty cents in my pocket and these two boys, who had
nothing at all, on my hands. Luck, however, was with
me. One of the first people I met was an old man named
Weeks, who ran a blacksmith and machine shop, and
when I showed him my credentials and told him my
qualifications he employed me.

This was on the 13th of May, 1883. It was just after
the mining boom in the Cœur d'Alene had burst; and
Spokane, being the nearest settlement, was the dumping
ground for the horde of disappointed and destitute men
who tramped, foot-sore and desperate, out of the mountains. In those days there was only a train a day each
way on the Northern Pacific—and not always that.
When a train steamed into the little station at Spokane,
a crowd, of all sorts and conditions, gathered to watch it;
oome meiely curious; some looking for an unwatched
brake-beam to ride away on; some spotting any arrivals
who appeared to have money, and who later on were not
any too secure against hold-ups.

While I was standing talking to Weeks in front of a
stable, which he also owned, we saw a man who had arrived by my train come out of the little hotel called the
San Francisco House, and walk down the street toward a
stone smoke-house, directly across from where we stood.
He had on a light overcoat and had his hands in his pock-
ets. It was nearly dusk, and just as he reached the smoke-
house a man jumped out from behind it and shoved a
pistol in his face. There was a shot, and the man with
the pistol fell headlong into the street. It turned out

afterward that the traveller had had a pistol, one of the old British bulldogs, in his overcoat pocket, and when the robber told him to throw up his hands, he had simply fired through his coat. The wounded man died in a few hours, after exonerating his slayer by saying, "He is all right, I'd have got him if he hadn't got me." This was my introduction to Spokane.

I lived in my tent for a year. I secured the contract to carry the United States mail between the railroad station and the post-office. There was only one small sack a day each way, and not always that (I remember that once there was not a bit of mail either way for thirty days), so that it was not much of a job, but it threw me in with Mr. Heath, the postmaster, and he, having a section of land that he had homesteaded, about a mile out of town and upon which no one would live because there was no water, offered me three lots if I would build a house and dig a well. The two boys I had had with me when I came had got work and left, and there was a carpenter who had struck town with no money and no work, and I boarded him for a year to build my house. I dug the well. I got my lumber from an old fellow who owned a saw-mill and needed some repairs done on his machinery, but had no money. I did the repairs and took my pay in lumber. Two years later I sold the house and lots for $2,750 cash.

Meanwhile I worked in Weeks's machine-shop, and one Saturday afternoon, late in August, having heard of a place where there were so many grizzly bears that no one dared to go there, I started out with two other fellows who thought they wanted to hunt bears, reached the promised land well after dark, and Sunday morning started on my first bear hunt. Looking back on it now, I think my idea must have been that hunting grizzly bears was something like "chumming" for fish; that all that was necessary was to go into the hills, let one's scent blow down breeze, and then shoot the ferocious animals that worked their way up wind with the intention of eating you.

At any rate our inexperience in this kind of hunting, our lack of caution, and our carelessness in making too much noise, prevented us from getting any grizzlies; and it spoke volumes for the number of these animals then roaming the hills that we actually, in spite of our awkwardness, saw eleven. I did succeed in killing one black bear, wounding another, and scaring several more nearly to death by rolling rocks after them. As a matter of fact it was well on into my second season before I really killed a grizzly, and although I saw a great many in my excursions, it was also about that long before I realized that the bear stories I had heard were just stories. I used to go out the last of every week.

Later, in the first fall, having saved some money, I bought a half-interest in the shop from Mr. Weeks. This I did really to have more freedom to hunt. After a partnership of five or six months, the old man gave me the other half of the business to teach one of his boys the trade, and I ran the shop till the next summer, when, the berries and grizzlies being ripe, I took a partner to look after the business while I looked after bear. Judging from the way he ran it, and the number of bear I got that year, I think it would have been cheaper for me to buy my bear. However, except for three or four months in the fall, when the hunting season was on, I ran this shop until 1888.

By that time the town had grown, I had bought property that had increased in value, I had got married, I had a home and an income, and I sold the blacksmith shop, devoted myself to hunting grizzlies and contracting for the building of houses in the town, and soon had most of my means invested in building materials. Then, in 1889, came the big Spokane fire. It burned the greater part of the city, wiped out my building materials, and as I carried no insurance, came near to wiping me out too; and what the fire only began, the hard times that followed completed.

I now found myself with an undiminished interest in hunting, and no means to gratify my inclinations. For a time I removed to Missoula, Montana, worked in a taxidermist's shop in the winter, and photographed and hunted during the summer; but I was soon to find a more congenial occupation. I was, it is needless to say, no longer the tyro of six years before. I had long since freed my mind from the preconceptions of lies and legends, and had worked out my own hunting lore from my own experience and observation. I was, it is true, familiar with only a section of the grizzly's range, and had penetrated but a few hundred miles in either direction from my base of operations. But I had always had strongly developed what some people call the bump of locality and direction, and what others more poetical refer to as "the sixth sense of the homing pigeon." I have, it seems to me now, always been at home in the woods. I have never been in any part of the country, in any mountains or any place, where it was difficult for me to find my way. I have never carried a compass, have never slept away from camp overnight, and have travelled three or four hundred miles across uninhabited districts and come out where I wanted to. I soon found that I had many friends who were anxious to hunt, and who were ready to pay me to take them hunting. These friends had friends. It was not long before I turned what had been my hobby into a business. Heretofore I had hunted three or four months a year. Henceforward I was seldom that long away from the woods. And soon I was familiar with every range of mountains in which the grizzly bear was found, from Mexico to Alaska.

In the beginning, I studied the grizzly in order to hunt him. I marked his haunts and his habits, I took notice of his likes and dislikes; I learned his indifferences and his fears; I spied upon the perfection of his senses and the limitations of his instincts, simply that I might the better slay him. For many a year, and in many a fastness of the hills, I pitted my shrewdness against his, and my wariness against his, and my endurance against his; and many a time I came out winner in the game, and many a time I owned myself the loser. And then at last my interest in my opponent grew to overshadow my interest in the game. I had studied the grizzly to hunt him. I came to hunt him in order to study him. I laid aside my rifle. It is twelve years since I have killed a grizzly. Yet in all those years there is not one but what I have spent some months in his company. And then (alas! that it had not been sooner) I undertook to photograph him. And finally I have attempted to put into this volume something of the story of the grizzly during the seventy-five years between his discovery and my meeting with him; something of my personal adventures with him; and the gist of my observations of his habits and of my opinions of his nature.

II

EARLY HISTORY—LEWIS AND CLARK

THE history of the grizzly bear differs from that of all the other great beasts that have come into close contact with civilization. The story of the others begins with our beginnings. The lion and the tiger have been always with us. They helped to rock the cradle of the race, and lunched occasionally from its contents. When we were cave men, we barred them from the mouths of our caves, and drew pictures of them on the walls. Later, we charred the ends of sharpened sticks in our fires, and with these drove them into the jungle. We and they have grown up together.

But the first chapter of the history of the grizzly is the beginning of the story's end. When my grandfather was born, the grizzly had never been heard of. If my grandson ever sees one, it will likely be in the bear pit of a zoölogical garden.

The actual history of the grizzly bear begins on April 29, 1805, when, on the banks of the upper Missouri, at the mouth of the Yellowstone River, in what is now Montana, Captain Meriwether Lewis, of the Lewis and Clark Expedition, met one of these animals for the first time.

Before this, indeed, hints and rumors of a bear different from the Eastern variety had come back to civilization

with returning traders and explorers. Edward Umfre-
ville, writing in 1790 upon "The Present State of Hudson's
Bay," had heard of them. In summing up the fauna of
the North and West, he says: "Bears are of three kinds:
the black, the red, and the grizzle bear." But he goes no
further than to add, in regard to the two latter, that, "their
nature is savage and ferocious, their power dangerous, and
their haunts to be guarded against."

Sir Alexander MacKenzie, the explorer, during his
second voyage, on May 13, 1795, mentions seeing bear
tracks on the banks of the Peace River, some of which
were nine inches wide. He says, "The Indians entertain
great apprehension of this kind of bear, which is called
the grisly bear, and they never venture to attack it except
in a party of at least three or four." He never, however,
seems to have seen one, nor does he describe it.

Lewis and Clark, on the other hand, not only entered
in their journals full accounts of their various encounters
with these animals, but made inquiries about them among
the inhabitants of the regions where they were found, and
took in them not only the interest of the fur trader and
the hunter, but that of the naturalist. Moreover, for
nearly fifty years these field notes of theirs were the chief,
if not the only, source of information regarding these
animals. Here and there, during that period, in the works
of scientific writers upon natural history, an original
observation, or an authenticated report of such observations, appears. But for the most part everything outside
the categories of frank romance or alleged adventure that
found its way into print, was an unblushing rehash or an
unwarranted elaboration of their statements.

Their note of April 29, 1805, is as follows:

"Captain Lewis, who was on shore with one hunter, met, about eight o'clock, two white bears. Of the strength and ferocity of this animal the Indians had given us dreadful accounts. They never attack him but in parties of six or eight persons, and even then are often defeated with a loss of one or more of their party. Having no weapons but bows and arrows, and the bad guns with which the traders supply them, they are obliged to approach very near to the bear. As no wound, except through the head or heart, is mortal, they frequently fall a sacrifice if they miss their aim. He rather attacks than avoids a man, and such is the terror which he has inspired, that the Indians who go in quest of him paint themselves and perform all the superstitious rites customary when they make war on a neighboring nation.

"Hitherto, those bears we had seen did not appear desirous of encountering us; but although to a skilful rifleman the danger is very much diminished, yet the white bear is a terrible animal. On approaching these two, both Captain Lewis and the hunter fired and each wounded a bear. One of them made his escape. The other turned upon Captain Lewis and pursued him seventy or eighty yards, but being badly wounded the bear could not run so fast as to prevent him from reloading his piece, which he again aimed at him, and a third shot from the hunter brought him to the ground. He was a male, not quite full-grown, and weighed about three hundred pounds. The legs are somewhat longer than those of the black bear and the talons and tusks much longer. Its color is a yellowish brown; the eyes are small, black, and

piercing; the front of the fore legs near the feet is usually
black, and the fur is finer, thicker, and deeper than that
of the black bear; add to which it is a more furious
animal and very remarkable for the wounds which it will
bear without dying."

Thus reads the first account of a meeting between a
white man and a grizzly.

I quote at length from Lewis and Clark on this subject,
not only because their notes are interesting, accurate, and
instructive in themselves, but because, while they are
scattered through the pages of a voluminous and unfamiliar report, a first-hand acquaintance with them is,
in their field, the beginning of knowledge. On May 6,
following, the record proceeds:

"Captain Clark and one of the hunters met this evening
the largest brown bear we have seen. As they fired he
did not attempt to attack, but fled with a most tremendous
roar; and such was his extraordinary tenacity of life that,
although five balls passed through his lungs, and he had
five other wounds, he swam more than half across the
river to a sand bar and survived twenty minutes. He
weighed between five hundred and six hundred pounds
at least, and measured at least eight feet seven and a
half inches from the nose to the extremity of the hind
feet, five feet ten and a half inches around the breast,
three feet eleven inches around the neck, one foot eleven
inches around the middle of the fore leg, and his talons,
five on each foot, were four and three-eighth inches in
length. This differs from the common black bear in having its talons longer and more blunt; its tail shorter; its
hair of a reddish or bay brown, longer, finer, and more

abundant; his liver, lungs, and heart much larger even in proportion to his size, the heart being equal to that of a large ox; his maw ten times larger.   Besides fish and flesh he feeds on roots and every kind of wild fruit."

May 11, 1805.   "About five in the afternoon one of our men (Bratton), who had been afflicted with boils and suffered to walk on shore, came running to the boats with loud cries and every symptom of terror and distress.   For some time after we had taken him on board he was so much out of breath as to be unable to describe the cause of his anxiety; but he at length told us that about a mile and a half below he had shot a brown bear, which had immediately turned and was in close pursuit of him; but the bear, being badly wounded, could not overtake him. Captain Lewis, with seven men, immediately went in search of him; having found his track, followed him by the blood for a mile, found him concealed in some thick brushwood and shot him with two balls through the skull. Though somewhat smaller than that killed a few days ago, he was a monstrous animal and a most terrible enemy.   Our man shot him through the centre of the lungs, yet he had pursued him furiously for half a mile, then returned more than twice that distance, and with his talons prepared himself a bed in the earth two feet deep and five feet long; he was perfectly alive when they found him, which was at least two hours after he had received the wound.   The wonderful power of life which these animals possess renders them dreadful; their very track in the mud or sand, which we have sometimes found eleven inches long and seven and one-fourth inches wide, exclusive of the talons, is alarming; and we had rather encounter two Indians than meet a single brown bear. There is no chance of killing them by a single shot unless the ball goes through the brain, and this is very difficult on account of two large muscles which cover the side of the forehead and the sharp projection of the centre of the frontal bone, which is also thick."

May 14, 1805. "Toward evening the men in the hindmost canoes discovered a large brown bear lying in the open grounds about three hundred paces from the river. Six of them, all good hunters, immediately went to attack him, and concealing themselves by a small eminence, came unperceived within forty paces of him. Four of the hunters now fired and each lodged a ball in his body, two of them directly through the lungs. The furious animal sprang up and ran open-mouthed upon them. As he came near, the two hunters who had reserved their fire gave him two wounds, one of which, breaking his shoulder, retarded his motion for a moment; but before they could reload he was so near that they were obliged to run to the river, and before they reached it he had almost overtaken them. Two jumped into the canoe, the other four separated, and concealing themselves in the willows, fired as fast as each could reload. They struck him several times, but instead of weakening the monster, each shot seemed only to direct him toward the hunter; till at last he pursued two of them so closely that they threw aside their guns and pouches and jumped down a perpendicular bank of twenty feet into the river. The bear sprang after them, and was within a few feet of the hindmost when one of the hunters on shore shot him in the head and finally killed him. They dragged him to the shore and found that eight balls had passed through him in different directions.   The bear was old and the meat tough, so that they took the skin only."

May 22, 1805. "We have not seen in this quarter [since passing the Muscle Shell] the black bear common in the United States, and on the lower parts of the Missouri, nor have we discerned any of their tracks. They may easily be distinguished by the shortness of the talons from the brown, grizzly, or white bear, all of which seem to be of the same family, which assumes those colors at different seasons of the year."

June 12, 1805. On coming out to the Missouri River from an expedition inland, they this day saw two large brown bears and killed them both at the first fire—"a circumstance which has never before occurred since we have seen that animal."

June 14, 1805, at the Falls of the Missouri. "Captain Lewis met a herd of at least one thousand buffalo, and being desirous of providing for supper, shot one of them. The animal immediately began to bleed, and Captain Lewis, who had forgotten to reload his rifle, was intently watching to see him fall, when he beheld a large brown bear, which was stealing on him unperceived and was already within twenty steps. In the first moment of surprise he lifted his rifle, but remembering instantly that it was not charged and that he had no time to reload, he felt that there was no safety but in flight. It was in the open, level plain—not a bush or a tree within three hundred yards, the bank of the river sloping and not more than three feet high, so that there was no possible mode of concealment. Captain Lewis therefore thought of retreating in a quick walk, as fast as the bear advanced, toward the nearest tree; but as soon as he turned, the bear ran open-mouthed and at full speed upon him. Captain Lewis ran about eighty yards, but finding that the animal gained on him fast, it flashed on his mind that, by getting into the water to such a depth that the bear would be obliged to attack him swimming, there was still some chance of his life; he therefore turned short, plunged into the river about waist deep, and facing about, presented the point of his espontoon. The bear arrived at the water's edge within twenty feet of him; but as soon as he put himself in this posture of defence, the bear seemed frightened, and wheeling about, retreated with as much precipitation as he had pursued. Very glad to be released from this danger, Captain Lewis returned to the shore, and observed him run with great speed, sometimes looking back as if he expected to be pursued, till he reached the woods. He could not conceive the cause of the sudden alarm of the bear, but congratulated himself on his escape when he saw his own track torn to pieces by the furious animal, and learned from the whole adventure never to suffer his rifle to be a moment unloaded."

June 20, 1805. "One of the men, who was sent a short distance from the camp to bring home some meat, was attacked by a white bear, closely pursued within forty paces of the camp, and narrowly escaped being caught. Captain Clark immediately went with three men in quest of the bear, which he was afraid might surprise another of the hunters who was out collecting the game. The bear was, however, too quick, for before Captain Clark could reach the man, the bear had attacked him and compelled him to take refuge in the water. He now ran off as they approached, and it being late, they deferred pursuing him till the next morning."

June 27, 1805. "As the men were hunting on the river, they saw a low ground covered with thick brushwood, where, from the tracks along shore, they thought a bear had probably taken refuge. They therefore landed without making a noise and climbed a tree about twenty feet above the ground. Having fixed themselves securely they raised a loud shout and a bear instantly rushed toward them. These animals never climb, and therefore, when he came to the tree and stopped to look at them, Drewyer shot him in the head. He proved to be the largest we had yet seen. His nose appeared to be like that of a common ox, his forefeet measured nine inches across, the hind feet were seven inches wide and eleven and three-quarters long, exclusive of the talons. One of these animals came within thirty yards of the camp last night and carried off some buffalo meat which we placed on a pole."

June 28, 1805. "The white bears have now become exceedingly troublesome, they constantly infest our camp during the night, and though they have not attacked us, as our dog, which patrols all night, gives us notice of their approach, yet we are obliged to sleep with our arms by our sides for fear of accident, and we cannot send one man alone to any distance, particularly if he has to pass through brushwood."

May 13, 1806, near the Kooskooskee River. "The hunters killed . . . a male and female bear, the first of which was large, fat, and of a bay color, the second, meagre, grizzly, and of smaller size. They were of the species common to the upper part of the Missouri and might well be termed the variegated bear, for they are found occasionally of a black, grizzly, brown, or red color. There is every reason to believe them to be of precisely the same species. Those of different colors are killed together, as in the case of these two, as we found a white and bay associated together on the Missouri; and some nearly white were seen in this neighborhood by the hunters. Indeed, it is not common to find any two bears of the same color, and if the difference in color were to constitute a distinction of species, the number would increase to almost twenty. Soon afterward the hunters killed a female bear with two cubs. The mother was black with a considerable intermixture of white hairs and a white spot on the breast, One of the cubs was jet black and the other of a light reddish-brown or bay color. The poil of these varegated bears is much finer, longer, and more abundant than that of the common black bear, but the most striking differences between them are that the former are larger and have longer tusks, and longer as well as blunter talons, that they prey more on other animals, that they lie neither so long nor so closely in winter quarters, and that they never climb trees, however closely pressed by the hunters. These variegated bears, though specifically the same with those we met on the Missouri, are by no means so ferocious, probably because the scarcity of game and habit of living on roots may have weaned them from the practice of attacking and devouring animals. Still, however, they are not so passive as the common black bear, which is also to be found here, for they have already fought with our hunters, though with less fury than those on the other side of the mountain."

May 31, 1806. (On one of the upper branches of the Columbia River.) "Two men visited the Indian village, where they purchased a dressed bearskin, of a uniform pale reddish-brown color, which the Indians called *yackah*, in contradistinction to *hohhost*, or white bear. This remark induced us to inquire more particularly into their opinions as to the several species of bears; we therefore produced all the skins of that animal which we had killed at this place, and also one very nearly white which we had purchased. The natives immediately classed the white, the deep and the pale grizzly-red, the grizzly dark brown, in short, all those with the extremities of the hair of a white or frosty color, without regard to the color of the ground of the poil, under the name of *hohhost*. They assured us that they were all of the same species with the white bear; that they associated together, had longer nails than the others, and never climb trees. On the other hand, the black skins, those which were black with a number of entirely white hairs intermixed, or with a white breast, the uniform bay, the brown, and light reddish-brown, were ranged under the class *yackah*, and were said to resemble each other in being smaller, in having shorter nails than the white bears, in climbing trees, and being so little vicious that they could be pursued with safety. This distinction of the Indians seems to be well founded, and we are inclined to believe: First, that the white, grizzly, etc., bear of this neighborhood forms a distinct species, which, moreover, is the same with that of the same color on the upper part of the Missouri, where the other species is not found; second, that the black, reddish-brown, etc., is a second species, equally distinct from the white bear of this country and from the black bear of the Atlantic and Pacific Oceans, which two last seemed to form only one species. The common black bear is indeed unknown in this country, for the bear of which we are speaking, though in most respects similar, differs from it in having much finer, thicker, and longer hair, with a greater proportion of fur mixed with it, and also in having a variety of colors, while the common black bear has no intermixture or change of color, but is of a uniform black."

It may here be noted that scientific naturalists, in their latest classifications of bears, while upholding Lewis and Clark in their surmises regarding the grizzlies, overrule them in their assumption that the other "black, brown, etc., bear" of the Rocky Mountain regions form a different species from the common black bear.

July 15, 1806. At the Falls of the Missouri again. "At night McNeal, who had been sent in the morning to examine the cache at the lower end of the portage, returned, but had been prevented from reaching that place by a singular adventure. Just as he arrived near Willow Run, he approached a thicket of brush, in which was a white bear, which he did not discover till he was within ten feet of him. His horse started and wheeled suddenly round, throwing McNeal almost immediately under the bear, which started up instantly. Finding the bear raising himself on his hind feet to attack him, he struck him on the head with the butt end of his musket; the blow was so violent that it broke the breach of the musket and knocked the bear to the ground. Before he recovered McNeal, seeing a willow tree close by, sprang up and there remained, while the bear closely guarded the foot of the tree until late in the afternoon. He then went off; McNeal being released came down, and having found his horse, which had strayed off to the distance of two miles, returned to camp."

III

FOLLOWERS OF LEWIS AND CLARK

ALTHOUGH the grizzly bear was thus discovered by Lewis and Clark in 1805, these explorers did not return to civilization until 1807, and even then their zoölogical records were not promptly worked up. It was not, therefore, until 1814 that any naturalist gave even tentative recognition to the discovery thus made. In that year, on the 4th of May, in his introductory address before the Literary and Philosophical Society of New York, Governor De Witt Clinton, speaking of the work which lay open to the coming naturalists of the country, referred to "the white, brown, or grizzly bear, the ferocious tyrant of the American woods," and spoke of it as "a nondescript, and a distinct animal from the polar bear, with which it is confounded." It ranged, he said, the country along the Missouri River where "it exists, the terror of the savages, the tyrant of all other animals, devouring alike man and beast and defying the attack of whole tribes of Indians." He then proceeded to call attention to the traditions of the Delaware and Mohican Indians, in regard to the Great Naked Bear, which they believed to have once inhabited the territories of New York State, and to suggest that the bear of this Indian legend was identical

with the grizzly. From another part of the transactions
of the Literary and Philosophical Society of New York,
I quote a curious version of this Indian legend. The report runs: "that among all animals that had been formerly in this country this was the most ferocious; that it
was the largest of the common bears and remarkably long
bodied. All over (except a spot of hair on its back of a
white color) naked. That it attacked and devoured man
and beast, and that a man or a common bear only served
for one meal for one of these animals; that with its teeth
it could crack the strongest bones; that it could not see
very well, but in discovering its prey by scent it exceeded
all other animals; that it pursued its prey with unremit-
ting ravenousness, and that there was no other way of
escape but by taking to a river, and either swimming
down the same or saving one's self by means of a canoe;
that its heart being remarkably small, it could seldom be
killed with an arrow; that the surest way of destroying
him was to break its backbone; that when a party went
out to destroy this animal they first took leave of their
friends and relations at home, considering themselves as
going on an expedition, perhaps never to return again;
that when out they sought for his track, carefully attend-
ing to the course the wind blew and endeavoring to keep
as near as possible to a river; that every man of the party
knew at what part of the body he was to take his aim;
that some were to strike at the backbone, some at the
head, and others at the heart; that the last of these animals
known of was on the east side of the Mohicanni Sipu
(Hudson's River) where, after devouring several Indians
that were tilling their ground, a resolute party, well pro-

vided with bows, arrows, etc., fell upon the following plan, in which they also succeeded, viz.: Knowing of a large, high rock, perpendicular on all sides and level on the top, in the neighborhood of where the naked bear kept, they made ladders (Indian ladders), and placing these at the rock, they reconnoitred the ground around, and soon finding a fresh track of the animal they hastily returned, getting on top of the rock and drawing the ladders up after them. They then set up a cry similar to that of a child, whereupon this animal made its way thither and attempted to climb the rock, the Indians pouring down their arrows in different directions, all the while upon him. The animal now grew very much enraged, biting with its teeth against the rock and attempting to tear it with its claws until at last they had conquered it."

The next mention made of the new bear occurs the following year, in 1815, in the second American edition of Guthrie's Geography where, upon information furnished by Brackenridge from the Lewis and Clark journal, George Ord, the naturalist, described and formally named the grizzly bear.

There were two words, similar in sound, but differing in signification, which had been impartially applied to this animal; one of them was *grisly*, and means "savage-looking, fear-inspiring, terrible, horrid"; the other was *grizzly*, and means "grayish, or somewhat gray." If one may judge from the context of Lewis and Clark's notes, they used the latter word with the latter meaning, but Ord evidently inclined to the belief that the first word had been used both with reason and intent, and he therefore gave to this species the name of *Ursus horribilis*.

Numerous other naturalists during the succeeding years suggested other names for the grizzly, such as *Ursus ferox* (De Witt Clinton), *Ursus candescens* (Hamilton Smith), *Ursus cinereus* (Richard Harlan). It being, however, a well-established rule that the name first given to a new species by its scientific classifier shall be retained, the grizzly bear is known to science as *Ursus horribilis;* and, it being also customary to add to the Latin name of any animal the name of the scientist who first formally described and named it, the Rocky Mountain grizzly is specifically known as *Ursus horribilis* Ord.

Mr. Ord, quoting Mr. Brackenridge as his authority, says:

"This animal is the monarch of the country which he inhabits. The African lion or the tiger of Bengal are not more terrible or fierce. He is the enemy of man and literally thirsts for human blood. So far from shunning, he seldom fails to attack and even to hunt him. The Indians make war upon these ferocious monsters with the same ceremonies as they do upon a tribe of their own species, and, in the recital of their victories, the death of one of them gives the warrior greater renown than the scalp of a human enemy. He possesses an amazing strength, and attacks without hesitation and tears to pieces the largest buffalo. The color is usually such as the name indicates, although there are varieties from black to silvery whiteness. In shape he differs from the common black bear in being proportionately more long and lank. He does not climb trees, a circumstance which enables hunters, when attacked, to make their escape."

Dr. D. B. Warden, in his "Account of the United States of North America," published in 1819, refers to the grizzly bear, but, apart from quoting some passages from Lewis and Clark, only adds that "the fur is employed for muffs and tippets, and the skins bring from twenty to fifty dollars."

D. W. Harmon, in his "Journal of Voyages and Travels in the Interior of North America," published in 1820, refers to the grizzly bear as "the gray bear," but does not, except in color, size, and strength, greatly distinguish them from "the brown or chocolate color, and those that are perfectly black." He refers to an occasional bear "the color of a white sheep."

In 1823 there was published "The Account of an Expedition from Pittsburgh to the Rocky Mountains Performed in the Years 1819 and 1820, by Order of the Honorable J. C. Calhoun, Secretary of War. Under Command of Major Stephen H. Long. Compiled by Edwin James from the Notes of Major Long, Mr. T. Say, and Other Gentlemen of the Exploring Party." Mr. T. Say, here mentioned, was the eminent naturalist, and his detailed scientific description and measurements of a half-grown grizzly shot by the party is printed in a foot-note on Page 53. They saw a number of grizzlies, but seem to have succeeded in giving them a wide berth. They met a number of Indians wearing necklaces and other ornaments made of the claws of the grizzly, and mentioned seeing and playing with a young bear of this species which was chained in the court of the Missouri Fur Company, near Engineer Cantonment. They quote freely from Lewis and Clark, and in their turn add several anecdotes, which are afterward repeatedly quoted by others.

Richard Harlan, in his "Fauna Americana," published in 1825, refers to the grizzly as *Ursus cinereus*, but does nothing more than quote from his predecessors.

The following year, however, in Godman's "Natural History," we come upon an interesting document. The author, after saying that the grizzly bear "slaughters indiscriminately every creature whose speed or artifice is not sufficient to place them beyond his reach," mentions two grizzly bear cubs that had, some time before, been kept alive in the menagerie of Peale's (afterward the Philadelphia) Museum. "When first received, they were quite small, but speedily gave indications of that ferocity for which this species is so remarkable. As they increased in size they became exceedingly dangerous, seizing and tearing to pieces every animal they could lay hold of, and expressing great eagerness to get at those accidentally brought within sight of their cage by grasping the iron bars with their paws and shaking them violently, to the great terror of spectators, who felt insecure while witnessing such displays of their strength. In one instance an unfortunate monkey was walking over the top of their cage, when the end of the chain which hung from his waist dropped through within reach of the bears; they immediately seized it, dragged the screaming animal through the narrow aperture, tore him limb from limb, and devoured his mangled carcass almost instantaneously. At another time a small monkey thrust his arm through an opening in the bear cage to reach after some object; one of them immediately seized him, and with a sudden jerk tore the whole arm and shoulder-blade from the body and devoured it before any one could interfere. They were still cubs and very little more than half grown when their ferocity became so alarming as to excite continual apprehension lest they should escape, and they were killed to prevent such an event."

He then quotes the following letter from Lieutenant Zebulon M. Pike, addressed to President Jefferson, referring to these cubs, and throwing a curious side light upon their ferocity.

WASHINGTON, *February* 3, 1808.

SIR: I had the honor of receiving your note last evening, and in reply to the inquiries of Mr. Peale can only give the following notes: The bears were taken by an Indian in the mountains which divide the large western branch of the Rio del Norte and some small rivers which discharge their waters into the east side of the Gulf of California, near the dividing line between the provinces of Biscay and Sonora. We . . . purchased them of the savages, and for three or four days I made my men carry them in their laps on horseback. As they would eat nothing but milk they were in danger of starving. I then had a cage prepared for both, which was carried on a mule, lashed between two packs, but always ordered them to be let out the moment we halted, and not shut up again before we were prepared to march. By this treatment, they became exceedingly docile, when at liberty following my men (whom they learned to distinguish from the Spanish dragoons by their always feeding them, and encamping with them) like dogs through our camps, the small villages, and forts where we halted. When well supplied with sustenance they would play like young puppies with each other and the soldiers, but the instant they were shut up and placed on the mule they became cross, as the jolting knocked them against each other and they were sometimes left exposed to the scorching heat of a vertical sun for days without food or a drop of water, in which case they would worry and tear each other, until nature was exhausted, and they could neither fight nor howl any longer. They will be one year old on the first of next month (March, 1808) and, as I am informed, they frequently arrive at the weight of eight hundred pounds.

While in the mountains we sometimes discovered them at a distance, but in no instance were we able to come up with one, which we eagerly sought and *that* being the most inclement season of the year, induces me to believe they seldom or never attack man unprovoked, but defend themselves courageously. An instance of this kind occurred in New Mexico, while I sojourned in that province: three of the natives attacked a bear with lances, two of whom he killed and wounded the third, before he fell the victim.

With sentiments of the highest respect and esteem,
Your obedient servant,
Z. M. PIKE.

HIS EXCELLENCY, THOMAS JEFFERSON,
  President of the United States.

Richardson's "Fauna Boreali Americani" (1829) mentions that a young grizzly cub caught in the Rocky Mountains was brought to England by the Hudson's Bay Company and kept alive in the Tower. Landseer made several engravings of it and a fine plate is included in Richardson's work. Here we first meet with the tale of a trapper seized and carried off from beside his camp-fire by a large grizzly, but rescued by a comrade. This occurred on the Saskatchewan, and Richardson met and talked with the rescuer, whose name was Bourasso, and who had an excellent reputation for veracity. The story is later quoted by Audubon. Theodore Roosevelt, in "The Wilderness Hunter," tells of meeting a French trapper named Baptiste Lamoche, whose head was twisted to one side from the bite of a grizzly bear which (according to his story) had sneaked up on him while sitting cooking dinner in camp by the shore of a lake, and had seized him by the neck with his teeth and started to drag him off into the woods, but was shot by one of his companions.

Audubon, in "The Viviparous Quadrupeds of North America" (1846), begins his article upon the grizzly bear by saying: "While in the neighborhood where the grizzly bear may possibly be hidden, the excited nerves will cause the heart's pulsations to quicken if but a startled ground-squirrel run past, the sharp click of the lock is heard and the rifle hastily thrown to the shoulder before a second of time has assured the hunter of the trifling cause of his emotion." Audubon himself, on August 22, 1843, had assisted in the killing of a grizzly bear on the Upper Missouri and his words are significant. They paint very clearly the frame of mind with which even a trained observer approached the study of this animal, and go far toward explaining why all the testimony relating to the grizzly bear's wariness and disinclination to fight unless pressed is uniformly overlooked by commentators, and only his ferocity dwelt upon. Audubon, for instance, goes on to cite Richardson's story of Bourasso, and the companion seized at the side of his camp-fire and made off with, and seems to regard it as quite what was to be expected. He then sites Drummond, the botanist, who, in 1826, in the Rockies, often came upon the grizzly bear unexpectedly, but said that when he stood still and watched them, or simply waved his hand, or made a noise with his tin box of specimens, they would rear up, look at him, and make off. He makes no comment upon this, however, and draws no inferences from it, and evidently never thought to try the experiment himself. One wishes, by the way, that Drummond had been a student of animals instead of plants. He had the right kind of stuff in him. He evidently came from Missouri, and he had great opportunities. For instance, in the latter part of June, 1826, he saw a male grizzly caressing a female. Soon after, both came toward him, whether by accident or to attack he did not wait to see, but climbed a tree. He then (being after all only a botanist) shot the female, and the enraged male rushed up to his tree and reared against it, but did not try to climb. He then returned to the female, which had sunk to the ground, and Drummond shot him too. So was wasted a chance to watch a forest courtship, to observe which I would tramp a hundred miles and live in a tree for a week.

So much then for the early history of the grizzly. It is not much, but it is all we have. Lewis and Clark's observations are the basis of it, repeated with slight variations and considerable embellishments in regard to ferociousness and bloodthirstiness by each after writer. Occasionally one of these adds an original observation or a hearsay anecdote. Then these in turn are repeated and embellished.

Meanwhile, the grizzly had been seized upon as a literary godsend in another quarter. To the romancers, the discovery of an *Ursus horribilis* was like the throwing open to settlement of a new territory, and there was a regular stampede to locate quarter sections. Captain

Mayne Reid was the hero of the movement. Jenkins, Lawrence, and a host of others preceded and followed him. Kit Carson wrote reliably and was not listened to. Jim Bridges told whoppers and was believed.

IV

JAMES CAPEN ADAMS

AND so we come to James Capen Adams. Adams was born in Medway, Mass., in October, 1807. He was trained as a shoemaker, but as soon as he attained his majority he joined a company of showmen as a collector of wild animals and hunted for them in the woods of Maine, New Hampshire, and Vermont. Later a tiger, belonging to the show, having disabled him while he was training it, he invested all his means in boots and shoes and started for St. Louis in search of health and a fortune. Finding neither in this outpost of civilization he joined the rush to California, where he arrived in the fall of 1849, having come overland *via* Mexico. Here for three years he engaged, with varying success, in mining, trading, and stock-raising, and finally becoming disgusted with the world and his fellows, in the fall of 1852 he took to the mountains and became a hermit, a hunter, and a purveyor of wild animals to shows and menageries.

At first he took no especial interest in grizzlies and, indeed, avoided them. He says: "I frequently saw him [the grizzly]; he was to be found, I knew, in the bushy gorges in all directions, and sometimes, in my hunts,

I would send a distant shot after him; but, as a general rule, during this first winter, I paid him the respect to keep out of his way; and he seemed somewhat ceremonious in return. Not by any means that he feared me; but he did not invite the combat, and I did not venture it." Later on he "considered it a point of honor to give battle in every case."

But had he been merely a hunter, merely even an uncouth knight-errant of the mountains, sworn to perpetual pursuit of the grizzly dragon, his story would not concern us. It was because he dealt in living grizzlies as well as dead ones; because for all his sworn enmity he admired, understood, and even loved them, and was the first white man to domesticate them; because, although he was neither a student nor even an educated man, he was yet, within the limits of his interest, an accurate observer, that I rank him so high as a light-giver on the subject of these animals. The story of Adams's career is told in a book called "The Adventures of James Capen Adams, Mountaineer and Grizzly Bear Hunter of California," written by Theodore H. Hittell, published in 1860, and long since out of print.[1] I have already told how the discovery of this book excited my interest in hunting, and in the grizzly; but some years ago, wishing to refresh my memory in regard to it, I obtained a copy only after much searching.

Adams, in this book, describes several of his expeditions; one undertaken in May, 1853, in company with a young Texan named Sykesey and two Indians, in the course of which he visited Washington and Oregon Territories, and after collecting many animals, including both

[1] Republished, 1911, by Charles Scribner's Sons, New York.

black and grizzly bears, took them to Portland and shipped them to the East; one undertaken in the early spring of 1854 to the Yosemite Valley; and one later in the same year across the Sierra Nevada and to Salt Lake City. These expeditions brought him into contact with the grizzly throughout the greater part of its range in what is now the United States, and he recognizes and comments upon the distinction between the grizzlies of California and those of the North. The grizzly, he says, is "the monarch of American beasts, and, in many respects, the most formidable animal in the world to be encountered. In comparison with the lion of Africa and the tiger of Asia, though these may exhibit more activity and bloodthirstiness, the grizzly is not second in courage, and excels them in power. Like the regions which he inhabits, there is a vastness in his strength which makes him a fit companion for the monster trees and rocks of the Sierras, and places him, if not the first, at least in the first rank of all quadrupeds."

Again he says: "There are several varieties of the grizzly bear; or, to speak more properly, perhaps, the species has a wide range, extending to the British possessions on the north, to New Mexico on the south, and from the eastern spurs of the Rocky Mountains to the Pacific Ocean. His size, general appearance, and character vary with the part of this great region in which he is found; for, although courageous and ferocious in the Rocky Mountains, he is there neither so large nor so terrible as in the Sierra Nevadas, where he attains his greatest size and strength. The grizzly of the Rocky Mountains seldom, if ever, reaches the weight of a thousand pounds;

the color of his hair is almost white; he is more disposed to attack men than the same species in other regions, and has often been known to follow upon a human track for several hours at a time. Among hunters he is known as the Rocky Mountain white bear, to distinguish him from other varieties. The Californian grizzly sometimes weighs as much as two thousand pounds. He is of a brown color, sprinkled with grayish hairs. When aroused he is, as has been said before, the most terrible of all animals in the world to encounter, but ordinarily will not attack man except under peculiar circumstances. The grizzly of Washington and Oregon Territories resembles the bear of California, with the exception that he rarely attains so large a size and has a browner coat. His hair is more disposed to curl and is thicker, owing to the greater coldness of the climate. He is not so savage, and can be hunted with greater safety than either the Californian or Rocky Mountain bear. In New Mexico the grizzly loses much of his strength and power, and upon the whole, is rather a timid and spiritless animal."

It was on his first expedition, somewhere in eastern Washington, that, having shot an old grizzly that was followed by two yearling cubs, and having, after many difficulties and repeated failures, captured the youngsters, Adams came into possession of Lady Washington, destined thenceforth to be his companion and servant. She was already old enough to resent the restriction of her liberty, and it was not until he had supplemented kindness with discipline that she accepted her new position in the scheme of life. "From that time to this," Adams says, "she has always been with me; and often has she shared my dangers and privations, borne my burdens, and partaken of my meals. The reader may be surprised to hear of a grizzly companion and friend, but Lady Washington has been both to me. He may hardly credit the accounts of my nestling up between her and the fire to keep both sides warm under the frosty skies of the mountains, but all this is true." The details of her training, the gradual augmentation of her liberty, the way in which she came to follow him to the hunt, and finally to consent (at first under protest) to bear the trophies of these joint expeditions back to camp on her back, makes fascinating reading, and Adams seems, naturally enough, to have valued her affection. But the following year her nose was put out of joint. During one of his hunts in the Yosemite Valley, in the spring of 1854, he located the winter quarters of a grizzly bear, from which the occupant had not yet emerged, and deciding, from the sounds that reached him in his careful reconnoitring, that the occupant was a female with young, he determined to watch for her appearance, kill her, and secure the cubs. The adventure proved a thrilling one, and at its conclusion he found himself in possession of a grizzly bear so small and helpless that he only succeeded in raising it by inducing a greyhound, that accompanied the party and had a young family of her own at the time, to adopt it in lieu of two out of her three offspring. She objected strenuously at first, but soon gave in gracefully, and Ben Franklin and his foster-brother grew up in amity, and continued to be sworn allies through life. Ben, having never known the world under any other guise, accepted it frankly as he found it. He not only did not have to unlearn the habits of the savage, but seems never to have developed them, at least not toward his master. He was never chained, slept for the most part in Adams's company, and when at last the ultimate test of allegiance was unexpectedly presented to him, he took sides unhesitatingly with his adopted master against his own relations. Adams, while accompanied by Ben Franklin, was attacked by a wounded grizzly. Ben instantly joined in the fight, and, though himself badly bitten, saved his master's life. From that time on he was the apple of Adams's eye, his inseparable companion, and of all living beings on earth the best beloved.

One is reminded of a quaint story, quoted by several of the early commentators, of a grizzly bear once domesticated by a tribe of northern Indians. On the occasion of a visit from members of another tribe, the bear's owner, for a joke, ordered the bear to get into one of the canoes belonging to the visitors. The bear obeyed, but the owner of the canoe, resenting the intrusion, struck him, and, "since the bear had come to be regarded as one of their family" by the hosts, the blow was the cause of an intertribal war.

Ben accompanied his master on several of his later trips, and more than once, suffering from blistered feet, limped after the outfit in improvised moccasins. On one occasion it was only by the most heroic devotion that Adams rescued him from the desert, where the bear had fallen exhausted, and lay, bending imploring eyes upon his master, as he left to search for water and help. We have no record of the manner of Ben's death, but one can well imagine, after reading Mr. Hittell's book, with what a heavy heart Adams must finally, in 1859, have sailed for home without him.

Adams also trained and, in a sense, domesticated, two other grizzlies. One of these, called "Funny Joe," he captured as a young cub on his expedition to Salt Lake; the other came to him in an entirely different way, as a result of the same trip. At one of their camps near the Emigrant Trail they thought it necessary to mount a guard at night. "The guard usually consisted of two persons, relieved at midnight by two others. The last guard on one particular night were Tuolumne and one of the Indians, who reported to me in the morning that a strange bear had entered camp, made the acquaintance of Lady Washington, and after a *tête-à-tête* of an hour or so had retired again in a very peaceful and orderly manner to the mountains from which he came. They had not called me because of my fatigue during the day, they said, and because the visitor had been so civil that they did not think it necessary to disturb me. I, however, directed that if such a case should occur again, they should not fail to let me know.

"The next night the visitor returned, and being informed of it, I got up. It was about midnight, but the moon was shining, so that we could easily see him approach the Lady, who was usually chained at night. I took my rifle with the intention of killing the beast, but, on second thought, concluded it would be more to our advantage to give him the freedom of the camp, and accordingly did not disturb him. He remained until dawn and then retired. On the occasion of his return the next night—for, like a royal lover, he was very attentive—

Gray advised that he should be killed, but I opposed the proposition, and, for aught I know, he still roams in his native haunts."

Here is one of the times when one could wish that Adams's interest had had a more scientific bias. It is only roughly that we are able to set from seven to nine months as the approximate time before Lady Washington gave birth to a male cub. Adams named him Frémont, but he seems to have done little credit to his romantic begetting and his noble parentage, either in intelligence or looks.

In the *American Naturalist* for May, 1886, under the title of "Domestication of the Grizzly Bear," John Dean Caton, LL.D., discusses Adams's adventures, describes his taming of Lady Washington, Ben Franklin, and Frémont, and says that at first he looked upon this book as an entertaining romance or at least as much embellished. But that, "upon inquiring in San Francisco, I met reliable persons who had known him well and had seen him passing through the streets of that city followed by a troupe of these monstrous grizzly bears, which paid not the least attention to the yelping dogs and the crowds of children which closely followed them, giving the most conclusive proof of the docility of the animals." This is the only reference I have ever seen made to Adams's book, and Mr. Caton's glimpse of him in the streets of San Francisco is interesting, and, if that were needed, confirmatory.

In 1907, having been informed by Dr. C. Hart Merriam that the author, Mr. Hittell, was still living in San Francisco, I wrote to him asking for some information about

A CAMP IN THE SIERRAS

the origin of the book and his recollection of Adams. In reply I received the following letter:

SAN FRANCISCO, *December* 15, 1907.
MR. W. H. WRIGHT.

Dear Sir: Your letter in relation to "The Adventures of James Capen Adams, etc.," has given me great pleasure. The book, unfortunately, was published in the exciting and excited days of 1860, just before the breaking out of the Civil War, and was never properly placed before the public; but it is gratifying to find that it did here and there reach readers who became interested in it. It is possible, and indeed, likely, that it will be republished, and, if so, it will contain a preface giving an account of how I became acquainted with Adams and came to write his story, and a postscript relating to his death and what became of his big bears, so far as known to me.

Your own work in hunting and studying the grizzly excites my lively interest, and particularly so as you say my book, to some extent at least, directed your attention to the subject. As to the questions you ask, or any other inquiries you may make, I will cheerfully give you all the information within my knowledge.

Ben Franklin, Adams's favorite bear, died in San Francisco; but as my papers are not at hand, I cannot be certain just now about the exact date. According to my recollection, it was in the summer of 1859. About the end of that year Adams went East, carrying his animals in a sailing vessel around Cape Horn, but without his finest specimen. I knew Ben Franklin well, often played with him, and on several occasions rode on his back. The picture of him and his master, given at the head of Part Second of my book, entitled "Adams and Ben Franklin," presents excellent portraits. Lady Washington and Samson were both, as I understand it, taken to New York and exhibited there by Adams under the auspices of Barnum. I do not know what became of them. Ben Franklin was caught, as a small cub, in the spring of '54; Lady Washington, as a larger cub in 1853, and Samson, as a large bear, in the winter of 1854-5; that is, if Adams was truthful in his statements to me, as I thought and still think he was. As to Samson's weight, my recollection is that Adams said he had had him weighed on a hay scales. His show bills in San Francisco gave fifteen hundred pounds as his weight and I never heard it disputed, but as he was doubtless the big bear exhibited in New York, it is possible that the exact weight could be ascertained there.

As to Adams's death, Barnum, in his autobiography, gives an account of it, and I know nothing more than he tells except a few items found in the newspaper at the time. My recollection is that he died in Massachusetts. There was a depression in his skull just above the forehead, which he said was caused by a blow from a bear in the early part of 1855, as related on pages 313 and 314 of my book. On his passage around Cape Horn, on his way to New York, he, according to report, had a fight with an ape or baboon, which tore the wound afresh, and, though it healed again sufficiently to enable him to go about and attend to business, he eventually died from the effects of it.

As to the comparative sizes of Ben Franklin and Lady Washington with Samson, I should say that the latter was nearly, if not quite, twice as large as either of the others; so far as I know he was the same bear that was exhibited in the Eastern States in 1860. He was untamed and had to be kept in a very large and strong cage, though I never saw him very wild.

In reference to getting Adams's story, I was in the newspaper business at the time, and could get only an hour or two a day to spend with him; and, as he talked, I wrote down what he said, usually in his own language, but sometimes with some changes to make it more grammatical. He knew little or nothing about the geography of the country, and I therefore could not locate him except in very general terms. He did not, on any occasion, appear to exaggerate, and told nothing improbable, though I had to wonder how he could remember so distinctly the particulars of his various hunts. I still have my notes of his talk. My object in writing the book was to tell his story in his own way, and I added nothing to the substance of his narrative except a few supposed embellishments and a little sentiment, besides literary order, expression, and arrangement. I have to thank Dr. Merriam for directing you to me, and hope my answers to your inquiries will be satisfactory.

Hoping to hear from you again, believe me,
Very sincerely yours,
THEODORE H. HITTELL.

To my mind, when I was a boy, this old man Adams was the prince of all hunters. Boone and Crockett and Carson seemed one-candle-power lamps to this old arc light of an Adams, and in some ways I feel so still. Adams, of course, was not a naturalist. He was not, except in his capacity as a hunter, trader, and trainer of wild animals, interested in natural history, that is to say, he was only interested in those habits and in those traits of the animals he dealt with that had to do with his success. But he was of a quicker intelligence and of a more independent nature than most of his kind; he insisted upon using his own eyes, he had a widely varied experience, and his reminiscences abound in observations of interest, and of at least conditional value. I shall more than once refer to them as occasions arise.

V

THE SCIENTIFIC CLASSIFICATION OF BEARS

IT was formerly the custom to class the North American bears in three groups—Blacks, Grizzlies, and the Polar Bear. The study during recent years of a series of several hundred skulls, including many belonging to the huge bears of the Alaskan coast region, showed this classification to be inadequate, and, scientific naturalists having added four more strongly marked species to our fauna, it became necessary, in view of the remarkable characters presented by these new forms, to rearrange our bears.

According to Dr. C. Hart Merriam's classification of the North American bears, these may now be classed in five well-marked, superspecific groups or types:

1. The Polar Bear.
2. The Black Bears.
3. The Grizzly Bears.
4. The Sitka Bear Type.
5. The Kodiak, or Alaska Peninsula Bear.

The five groups are unequally related. The Polar Bear belongs to an independent genus. The Black Bears differ more from the others, taken collectively, than the latter do from one another; and seem to be the only ones whose distinctive character is of sufficient weight to entitle them to subgeneric recognition.

Taking up these groups in order we find that:

1. The Polar or Ice Bear, *Thalarctos Maritimus* Linn., inhabits the Arctic shores and the islands of both continents, and has not been subdivided.

2. The Black Bears may be separated into at least four species, having, respectively, more or less circumscribed geographic ranges:

(*a*) The Common Black Bear, *Ursus Americanus* Pallas.

(*b*) The Louisiana Bear, *Ursus Luteolus* Griffith.

(*c*) The Florida Bear, *Ursus Floridanus* Merriam.

(*d*) The St. Elias Bear, *Ursus Emmonsi* Dall.

Some of these may be found to intergrade, and *Ursus Americanus* may be still further split into subspecies.

3. The Grizzly Bears (including the Barren Ground Bear) may be separated into four more or less well-marked forms, as follows.

(*a*) The True Grizzly, *Ursus Horribilis* Ord, from the northern Rocky Mountains.

(*b*) The Sonora Grizzly, *Ursus Horribilis Horriæus* Baird, probably only a subspecies.

(*c*) The Norton Sound, Alaska, Grizzly, probably another subspecies.

(*d*) The very distinct Barren Ground Bear, *Ursus Richardsoni* Mayne Reid.

Whether or not the large grizzly from Southern California deserves subspecific separation from the Sonora animal (*Horriæus*), has not been determined.

4. In the fourth group, the large brown bear of Sitka and the neighboring islands (and perhaps the adjacent mainland also), *Ursus Sitkensis* Merriam, and the large

brown bear of Yakutat Bay and the coastal slope of the St. Elias range, *Ursus Dalli* Merriam, are the representatives of a very distinct type. They resemble the grizzly in the flatness of their skulls, but are much larger, are different in color, have more curved foreclaws, and the Sitka bear has a different type of sectorial tooth. The Yakutat Bear is much larger than the Sitka Bear, and has very different teeth. It may represent a different section.

5. The gigantic fish-eating bear of Kodiak Island and the Alaska Peninsula, *Ursus Middendorffi* Merriam, is the largest of living bears and differs markedly from all other American species. It closely resembles the Great Brown Bear of Kamchatka, *Ursus Beringiana* Middendorff, which it slightly exceeds in size. The extraordinary elevation and narrowness of the forehead suffice to distinguish this bear from all other known species.

The number of full species of North American bears here recognized is ten: four of the Black Bear group, two of the Grizzly group, three of the big brown bears of Alaska, and the Polar Bear.

These ten species are as follows:

The Polar Bear, *Thalarctos Maritimus* Linn.

The Common Black Bear, *Ursus Americanus* Pallas.

The Louisiana Black Bear, *Ursus Luteolus* Griffith.

The Florida Black Bear, *Ursus Floridanus* Merriam.

The St. Elias Black Bear, *Ursus Emmonsi* Dall.

The Rocky Mountain Grizzly, *Ursus Horribilis* Ord.

The Barren Ground Grizzly, *Ursus Richardsoni* Mayne Reid.

The Sitka Brown Bear, *Ursus Sitkensis* Merriam.

The Yakutat Brown Bear, *Ursus Dalli* Merriam.

The Kodiak Brown Bear, *Ursus Middendorffi* Merriam.

It will, of course, be understood that when in this work mention is made of the grizzly bear, it is intended to refer to that species found throughout the northern Rocky Mountain region of the United States and British Columbia, the species recognized by all scientific naturalists as the true grizzly, *Ursus Horribilis* Ord. No other type of the grizzly will be discussed, except by way of comparison.

PART II

MY EXPERIENCES AND ADVENTURES

VI

MY FIRST GRIZZLY

HUNTING grizzlies requires more quiet, more skill, and more patience than any other kind of hunting, and at the best there will be more disappointments than killings. It took me several years to learn this. I started out with deep-rooted but mistaken notions, and these had to be knocked out of my head by failures, and new ones hammered in by experience, before I understood the grizzly well enough to even occasionally get the best of him.

The best way to hunt them is to study their habits, familiarize yourself with their range, and lie in wait for them near their feeding grounds. Trailing them is more than uncertain. It is, to be sure, a supreme test of woodcraft and endurance, but it must be a good hunter indeed that can take up the trail of an old grizzly and hope to get a shot at him.

I am often asked what is the best gun for grizzlies? My answer is that the best gun is the one the hunter is most used to and hence has the most confidence in; any of the high-power guns are all right and will, if the bullet be well placed, end matters at the first shot. My own first hunting rifle was an old .44 Winchester that I brought from the East. It had already done good service for deer and black bear, but the extractor had become worn, so that it would not always draw the shell unless I put my thumb on it and bore down. This worked satisfactorily as long as I thought to do it, but there were times when I forgot and then there was trouble. The cartridge then had to be cut to pieces before it could be taken out, and it required a pocket-knife and some labor to accomplish the result. For a year or more I had been thinking of getting another gun, but this one was so accurate that I clung to it, and at last it got me into difficulties.

I made a number of trips after grizzlies and I had got sight of several, but they had always seen or heard me first, and when I would see them they were just disappearing over some ridge or into some jungle. So one spring, having made up my mind to go after them and not return until I had one, I started out in May with a few pack-horses and went to the Bitter Root Mountains, which form the dividing line between Montana and Idaho. A friend went along with me to look after the horses, help do the packing and, as he said, skin the bears.

For nearly three months we cruised about this rugged wilderness and enjoyed life to the utmost. We killed plenty of black bears, but up to September had not bagged a grizzly. We found an abundance of their tracks and saw three bears, but they were so wild that we could not get near enough to them for a shot, and twice, when it seemed as if one could not possibly escape, it quietly slipped out of sight at a point I had not calculated upon.

We had, during this time, killed several deer for meat, but though we had seen some elk and one or two moose, we had not shot them since we could not care for so much meat and had no way of carrying the heads. At last, however, having, perforce, given up getting a grizzly, we turned our faces homeward, and I determined to kill an elk and pack out the head and what meat we could.

We therefore left the divide we had been following and struck off to the right to reach a stream of considerable size flowing into the main north fork of the Clearwater River. We had been told by an old miner that there was a large lick on this stream about twenty miles from the trail, and he directed us as to where to leave the ridge, and where, after we struck the stream, we would find the lick.

We followed his directions and arrived at the creek about noon the second day. The stream was very swift and cold, since we were near its source, and it flowed from the snow-banks only a few miles away. We set up our tent, turned the horses out to graze, had a quick lunch, and I, taking my rifle, went to search for the lick, leaving my friend to attend to the camp. The weather was warm, and expecting to return before dark, I did not take my coat. I found the lick where the miner had stated and saw plenty of fresh tracks of elk and deer. There was no timber nearer the spot than across the creek, except some two hundred yards down the stream, but close to the lick, on its lower side, there were some small bushes three or four feet high. Into these, then, I crawled, and finding an old log that had been left there by high water years before, I cut some of the bushes and made a rough bed on the lower side of it.

This was my first experience in watching such a place, and I did not know that, at that time of year, the game did not come every day. Judging from the trails, I took it that they did. They had, without a doubt, been there that morning, and so I curled down and prepared to wait. At first it was not uncomfortable. The sun shone in upon me and I found myself dozing several times. Later, when the sun sank behind the mountains, I began to feel the need of a coat, but still I did not care to give up, and determined to stick it out until I got an elk or it became too dark to see.

Every few minutes I raised my head above the bushes, but nothing seemed to be stirring. Finally it became so cold that my teeth began to chatter and I looked at my watch and made up my mind to stay just five minutes more. When these were gone I raised up, took another look, and seeing nothing in sight, concluded to remain another five. This I did several times. Finally I was nearly frozen and I determined that the next five minutes should be the last and that I would not wait longer for all the elk in the country. When the time was up I took a good look about, but not a living thing could be seen. Then I looked up-stream and, just around the point of the hill and out in the bottom about a hundred yards away, I saw what seemed to me to be the old grizzly I had seen in the cage when I was a boy. He had the same carriage, had the same big forearms and the gait I would know again anywhere as long as I lived. Best of all the brute was headed straight for my log. I ducked back to my bed and waited almost breathless for him to get nearer. I had laid no plans. I simply wanted him to get near enough that there might be no possibility of his escaping. It now seemed certain that my dream was to come true.

From a photograph, copyright, 1909, by J. B. Kerfoot

A THOUSAND-POUNDER, HALF AS BROAD AS HE WAS LONG

I waited for what seemed an unconscionable time, and hearing nothing, began to fear that this bear, like the others, had given me the slip. Finally I raised up to look about and there he was about seventy-five yards away. He had evidently been standing still and had resumed his walk just as I looked up. I did not drop down again, but just stooped so that he would not see me, and waited for him to reach a spot from which he could not get away once I had opened up on him.

When some forty yards off he turned a little to the left so as to avoid some bushes, and this being just what I wanted, I straightened up, aimed carefully just on the point of the shoulder close to the neck, and pulled the trigger.

It never entered my thoughts but that the bear would drop in his tracks. One can, therefore, imagine my surprise when he gave a roar like a mad bull and came my way on the jump. I threw the lever of the rifle forward to pump up another cartridge, but that roar, and the grizzly's tactics, had so surprised me that I forgot to thumb my old Winchester and the shell was left jammed in the breach. Here was a situation not down on the programme, and it began to look as though I was going to have all the grizzly I wanted. I took one look, dropped the gun, and lit out for the creek, and as I reached it I jumped down the bank, which was about three feet high.

Underneath this bank the dirt had been washed out by the river, leaving a considerable hollow below the roots of the bushes, and when I saw this space I ducked into it, figuring that the bear, if he jumped after me, might jump clear over me and give me a chance to climb back up the bank and make for the timber down the stream. The water was ice cold and I had been nearly frozen before taking to it, but I had no regrets. I waited for what seemed to me a half-hour, but I could hear nothing beyond the rush of the water as it surged around the boulders, and at last it got so cold that I felt that I would as soon be clawed by a bear as frozen to death. So I tried to make my way up-stream, thinking to reach a point above the bend and there get out and go to camp; but the water was so rapid and seemed to me to make so much noise as it struck against my neck, that I was afraid the bear would hear me if he was listening, as I supposed he was. He was behaving so differently from the bears I was accustomed to that I was at a loss how to size him up.

When I found I could not go up the stream I decided to go down, and by hanging on to the roots, slid noiselessly along for about fifty yards. Then, as I still heard nothing of the bear, I concluded that he had either left the spot or was keeping quiet and watching for me to reappear, and, determined to at least have a look, I got up on my knees, scraped the water out of my shirt, and peered cautiously over the bank. But I was still too low to see over the bushes, so I crawled ashore and raised up enough to look about me. Still I could see no bear. I now began to fear that he had escaped me, instead of I him, and I made up my mind to creep to my gun, cut the cartridge out, and hunt for him.

It makes me laugh to this day when I think of the picture I must have made, first crawling a few feet, then lying still and listening. It was the most conscientious stalking I ever did. When I reached the log and looked for my gun I could, at first, see nothing of it. But, raising my head a bit, I spied it lying where I had thrown it in my haste to get away, and I at once cut the shell out, carefully thumbed it while I pumped in another cartridge, and then, my own man again, I stood up and looked about me.

And the first thing I saw was that terrible bear, as dead as a stone, and not more than twenty feet from where it had stood when I shot it. Moreover, it had turned a complete somersault and was headed the other way. The bullet had entered between the shoulder and the neck, had hit no bones except a rib, and had passed over the heart, severing the large blood-vessels.

Since then I have found that nearly all grizzlies, if shot when they are not aware of the presence of the hunter, will, for some reason, run in the direction from which the wound is received. This, I believe myself, is why so many claim that grizzlies always charge them. The next time my attention was drawn to this I shot across a ravine at a grizzly that had just come out of the bushes. I obtained a side shot at about a hundred and fifty yards, and I made it sitting down, aiming close up to the shoulder. At the crack of the gun the bear turned and charged straight at me, and I could see the brush in the gully sway, and could hear him tearing through it. I ran back a few yards from the edge of the brush so as to get a clear shot at him when he came out of the ravine, and when he emerged I made a movement to shoot. Like a flash he saw me, evidently for the first time, and turned so quickly that, as I was not expecting such a move, I did not have time to fire. He only made a few jumps after seeing me and changing his course, before he fell dead.

Since then I have seen many bears do the same thing, and I am convinced that, although they generally, in these circumstances, run toward one, it is not by any means always with the intention of charging.

VII

FIVE IN FIVE SHOTS

I SHALL never forget the first grizzly-bear trail I ever saw. Not only because it was the first, but because for many months it and its maker interested me, cost me untold exertions, brought me uncounted disappointments, and finally figured in the most successful encounter I ever had with grizzlies.

It was a very distinct trail across the bottom of a cañon, perhaps a hundred yards wide, and had been worn by an old and very large animal. The ground on either side of the gully was too hard to show his footprints, but, coming and going, he always seemed to cross at the same spot, and I thought it would be an easy matter to watch the trail and shoot the bear. This was when my first greenness was a thing of the past and I was beginning to take notice on my own account; but after watching off and on for a week and over, and seeing nothing, I concluded that the bear no longer used the trail, and dismissed the matter from my mind. A few weeks later, however, in passing the spot again, I saw fresh tracks, and knowing that the bear was still in the vicinity, resumed my watching, but without result. Then I tried still-hunting the cañon. I crawled through every thicket and looked into every place where I thought a bear could hide, but aside from a fourteen-inch track and a few hairs that he would leave sticking to trees as he passed, I saw nothing that looked like a bear, and was almost willing to swear that no grizzly inhabited the cañon.

For two years I continued at intervals to see the big tracks in this cañon, but not once did I see the bear. Finally, both my knowledge and experience having been augmented, I made up my mind that, if such a thing were possible, I would at least set eyes on him, and I got a man to go with me to look after the camp and horses so I would have nothing to do but hunt. We made camp about two miles from the cañon, and my first excursion showed me that my old friend with the big feet was still in evidence.

The upper edges of this cañon were heavily timbered, and above this timber on one hand stretched an open hill-side facing the south. Near the head of the cañon this hillside was cut into by many little ravines, and along the edges between these the sarvis berry grew. As I had never succeeded in getting sight of the bear in the cañon itself, I decided to watch the hillside and perhaps catch him as he came out to feed; so I selected a point which commanded a view of the whole hill, and every morning from daylight until ten o'clock found me on the lookout, seated in a little clump of fir-trees. And from three o'clock until dark I was in the same place. Day after day, however, passed and brought no bear, and at last the camp tender, while he did not say out and out that he thought I was "locoed," intimated it pretty broadly. For his part, he said, he did not believe there was a bear in the whole country.

Yet, almost every morning, examination showed fresh bear tracks on the old trail, and I felt it safe to assume that something was making them. So for a whole week I lay in that clump of firs. Then I began to think about giving up; but, knowing that the animal *must* show himself in time, I kept taking on fresh stocks of patience and dragging myself again and again up to the little clump of firs.

At last there came a rain. It began in the night and kept up until about one o'clock in the afternoon of the following day, and rained so hard that I did not go out in the morning. In the afternoon it was so wet that I hesitated about going, but finally, thinking that this might be just the time the old bear would select to go berrying, I decided to risk it. And, as luck would have it, when I reached the firs and took a look at the hillside, there sat an old grizzly about a hundred yards above the brink of the cañon, and some three hundred yards from me, busily engaged in pulling down branches and eating berries.

I immediately began the sneak of my life. I did not, even at first, think of walking. I simply got down on the ground and snaked it. I worked below the bear, so that if he ran he would have to come my way or go up the open hillside and thus afford me additional shots if I failed with the first. But I had no intention of failing. I worked along slowly, so that the bear had moved quite a distance up the hill before I finally got within reasonable range, and even then I kept on until I was within sixty yards before finally making up my mind to risk a shot. I then crawled behind a bunch of bushes and, without getting up, looked the ground over to see what the chances were of the bear's getting back into the cañon in case I failed to drop him.   And down there, in the cañon I had watched so long and so vainly, stood the largest bear it had ever been my good fortune to set eyes on.   I began to think the woods were full of them, and backing silently into one of the small ravines, I worked down the hill toward the big fellow.   And as I got a better view of him I knew what had made those tracks.   I had thought nothing about the size of the first bear.   I had been disappointed so often that anything went.   But now that I had set my eyes on this big one I thought him entitled to precedence.

Yet I wanted both, and I thought I saw my way to getting them.   The first bear seemed to have struck a bonanza berry patch and was moving slowly, or not at all. The big fellow, on the other hand, was down where the bushes were pretty well stripped and seemed to be working uphill fairly fast.   I therefore dropped out of sight, wormed my way downhill a bit farther, waited till the two bears were about a hundred yards apart, and then crept to the top of a slight ridge and found myself some forty yards from the big one and sixty from the other.

I figured on killing the large bear at the first shot and then turning on the other before he had time to take in the situation; and I relied on the second bear's standing up to take a look before making for the cañon, and thereby giving me the few seconds that I would need.

By this time I had got rid of my old .44 repeater, and was shooting a single-shot .45-100 Winchester that weighed twelve pounds.   I had selected this gun because I could always depend on it.   I used the full charge of

From a photograph, copyright, 1909, by J. B. Kerfoot

GRIZZLIES FEEDING

powder, and had had swages made to swage slugs that would weigh six hundred grains of soft lead. One of these placed in the centre of a grizzly's shoulder never failed to decide matters.

I took a sitting position that afforded me a right-hand quartering shot at the big fellow and a left-hand quartering shot at the other, and that would enable me to act very quickly after the first shot. For this I intended to take my time, and to trust to luck and rapid work for the second. And with one cartridge in the rifle, three others on the ground, and two more between the fingers of my right hand, the old single-shot could be depended upon for both.

When all these things were arranged to my liking I waited for a side shot at the large bear. I did not have long to wait, and I never looked through sights more carefully than when drawing that bead. There were only about two square inches of bear visible when I pulled the trigger, but they were the exact square inches that I wanted, and once the shot was delivered I wasted no time in finding out the result but turned to the other bear. He acted exactly as I had expected. He turned side on to me in order to see what bedlam had broken loose. I caught him squarely in the shoulder, and he wilted in his tracks as the other had done. Not a yard did either of them move after being shot.

I now got up to examine my prizes when I heard a clawing and rolling of gravel in the next ravine. Glancing about to see what caused this racket I faced, to my intense surprise, a mother bear and two half-grown cubs, their retreat to the cañon having been cut off, making for the top of the divide as fast as they could go. I dropped back to my sitting position, with my elbows on my knees, and took a flying shot at the old bear. She was not over seventy-five yards away and must have been quite near me, in the next ravine, when the firing began. My bullet caught her with a quartering rake forward, and rolled her back into the gully, and as this, of course, stopped the cubs, they fell to the next two shots.

I was soaked to the skin from the wet brush, and plastered from head to foot with mud and dirt. But that, and the endless waiting and watching by the clump of firs, yes, and all the disappointments that had gone before, were paid for now. Five grizzlies down to as many shots, in as many minutes, cancels many debts. This was the greatest bag of grizzlies that I ever made single-handed.

VIII

GRIZZLY GOURMETS

THERE is an old saying that the way to a man's heart is through his stomach. This, in another sense, is equally true of bears; and more grizzly hunters have won their chance to drive their bullets home by studying their victim's appetite than by any other method. It follows that some of the easiest hunting grounds in the north-west used to be along the streams where the salmon ran, for the grizzly is a great fisherman. It is true that in the fishing season his pelt is valueless, as at that time of year he has no fur and but very little hair, but the man who has come out to get a grizzly is apt to look upon this circumstance as, indeed, a misfortune, but one to be taken philosophically.

In the streams tributary to the Clearwater River in Idaho there are two or three runs of salmon. One, of what are locally known as the red, or Columbia River salmon, takes place in the early spring, at the time of high and muddy water. It does not, on this account, attract so much attention from the bears. But later on, between the middle of August and the middle of September, what are known as the dog salmon make their way up all the little streams. At that time the water is clear and low; the stream beds are successions of small rapids and broken riffles; the mad, unreasoning longing of the salmon to reach the uttermost head of salmon navigation drives them to struggle over places where there is scarcely enough water to float them when swimming on their sides, and the grizzlies gather to the feast.

These dog salmon grow to a very large size. They run from two to four feet in length, and would, if fat, weigh fifty or sixty pounds. But by the time they reach the upper waters of the small streams they are very poor and thin (good, indeed, only for bear bait), and fall easy victims to the bear and other animals and large birds that prey on them.

The grizzly has his own calendar and never gets mixed up on it. About two weeks before the salmon are due he leaves the higher hills and ridges and gathers near the streams, to be on hand when the fish appear. One can easily tell where to lie in wait for them. They always have their favorite fishing ground, usually at some shallow riffle, and the creek bank is worn smooth by the many trails leading away into the dense thickets, where they lie up when not fishing. After locating one of these fishing grounds, one has only to select a good hiding place, being careful not to make noise enough to frighten away the bears, and then wait until they come down to fish. In localities where they have been little hunted, I have seen them out fishing at all hours of the day, and to see a grizzly catching salmon is worth one's while.

The grizzly usually sits on the bank of the stream and watches the riffles over which the salmon try to force their way. He will wait quietly enough until the salmon is about half-way up the riffle and struggling in its efforts to make the ascent. Then he will make a quick dash, and, with one sweep of his huge paw, will send a shower of water ten feet into the air, in the midst of which will be seen a salmon sailing toward the creek bank and landing, many times, ten or twenty feet beyond. Then the bear hurriedly makes for the shore and, if hungry, eats the fish. If he has already had his fill, he will kill it, lay it down, and, returning, wait for another. I have seen one bear catch seventeen salmon in this manner before stopping, and he then carefully piled them together and buried them for future use.

Sometimes a bear will sit on a log jam and watch for the fish to swim out from under the logs. When one comes he will, with a sweep of his paw, send it flying to the bank. I have often seen them fishing in this way, lying on a log with one paw hanging in the water, and it is wonderful how many salmon they will fling out. Once I saw five old grizzlies fishing from one log jam. Indeed, I have watched for hours along these streams, and some of the pleasantest moments of my hunting trips have been so spent; but while I could have killed many a bear in this way, I have never killed but three.

When I first began to hunt, I thought that the salmon run would be a good opportunity to make a fine score in bears, and accordingly, on one of my trips, having found a place that was all tracked and worn by grizzlies, and where there were large piles of decayed, ill-smelling salmon, I stationed myself a little above the riffle and waited. About three o'clock in the afternoon an old bear came solemnly out of the thicket to the edge of the stream, sniffed about for a while, and then walked out on a log that lay across the creek just below, or at the foot of, a riffle, where there was quite a long and deep hole. In such holes the salmon sometimes collect until there are as many as fifty. Then, all together, they will begin to climb the riffle, and for an hour it is "Step lively, please," all along the line. Soon after this bear walked out on the log, another one made its appearance and it, too, walked out on the log and sat there watching the salmon in the hole below. I had never, then, seen bears catch salmon, and so, for the time being, refrained from shooting. The fish, however, did not seem to be in any hurry to get up the stream, and I waited for an hour or more before anything happened. Finally I saw one of the bears gather himself, look eagerly into the stream, and move his head as if following something that was stirring there. Then I saw the back of a salmon on the riffle, and the water began to boil as the fish tried to force its way through the shallows. When it was some feet above the log, and about half-way over the riffle, one of the bears gave a spring and a stroke of its paw, and the trick was done. The salmon was hurled through the air to the bank and out of sight in the brush, where the bear followed it. The other bear was now in the act of grabbing a salmon, and he was equally successful. By that time the first one was back again. There was no loafing. Those bears were paid by the piece and seemed to know it. I was altogether too interested to think of shooting, but sat, open-mouthed, watching the finest fishing I had ever seen.

The racket on the riffle soon brought another bear out from the brush, and before the show closed there were

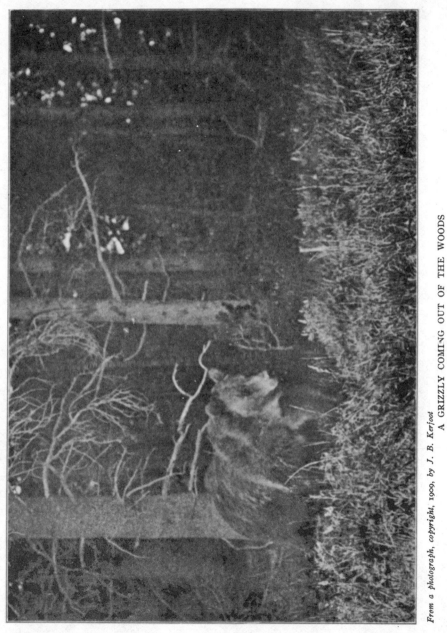

From a photograph, copyright, 1909, by J. B. Kerfoot

A GRIZZLY COMING OUT OF THE WOODS

four of them engaged in the sport. They must have caught fifteen or twenty before there was any let up. Then the salmon did not come so fast, and one bear, on going to kill his catch, forgot to come back. The other three were evidently becoming restless, as if they had had about enough for the time being, and I thought it high time to begin shooting if I meant to do any.

I was then using my single-shot rifle, made to order for me by the Winchester people: the .45-100, in which I shot one hundred grains of powder and six hundred grains of lead. It was one of the guns that killed at both ends, but I liked it better than any I have ever carried. I used it for years, and I discarded it for a lighter .30-30 only when I gave up hunting with a gun and took to hunting with a camera. Personally, I could depend on this old rifle for a sure three shots in twelve seconds, by holding two spare cartridges between the fingers of my right hand, and I have always thought that a hunter is apt to be much more careful if he knows that every shot must tell. I always got as close to the game as I could before shooting, and whatever I shot, it generally dropped, if hit, and I was usually near enough to be sure of hitting.

I presume that, as I looked at the three remaining bears, I was the victim of what nowadays would be called a game-hog feeling. At any rate, I found myself figuring on how I might get them all before they could get out of reach. I was a little up-stream from them, and not over fifty yards away. I figured that they would not come toward my side of the creek when I fired, and that if I killed the one nearest to the other shore, its body, if remained on the log, would retard the others in their effort to escape. There was room for but two of them on the log at one time, and the third was now on my bank. The largest was in the middle, the other two being about of a size. I made up my mind, therefore, to first chance a shot at the small bear on the log and then to turn my attention to the largest one. Afterward, if things worked out as I hoped, I could pay my compliments to the one on shore.

I accordingly fired at the small bear, and hitting him square in the shoulder, he dropped promptly and without a murmur, but he dropped into the pool. Before the other one on the log had time to make much of an investigation I hit him near the same place, but a little further back, and he made for the shore. I paid no attention to him, as I knew that he was fatally wounded and would not go far, and slapping my third cartridge in place, I turned to look for the third bear, but all I could see was the swaying of the bushes where he had disappeared. The second one shot went some fifty yards after reaching the bank, when he fell, and was quite dead when I got to him.

But now that the excitement was over and I had time to take stock of my achievement, my satisfaction was short-lived. The hides were not worth taking off. It had been a useless slaughter and I was sorry that I had killed. I took the large teeth and long claws of the dead bears, but since then I have never, but once, shot at a grizzly when it was fishing.

This was some years later. I was out hunting with a man from New York who was very anxious to kill a grizzly, having never shot one; so we made our way to one of the creeks, where the bears fish, and soon finding a suitable place, began our watch. This was some two miles below our camp, where there was quite a wide bottom, through which the creek ran. There were several channels in the stream that had been formed by log-jam obstructions during high water, and among these channels were several islands. It was an ideal fishing ground, but, as the bears worked all the channels, it was an even guess as to just where we would catch them. For a couple of days we watched one place without success, and on the second evening were returning to camp, after watching until dark, and were crossing one of these islands on a foot log. Just as we passed around the large root at the butt of a log there was a snort and a splash that, expecting nothing of the sort, startled us and nearly threw us off our feet. The commotion was not over ten feet from us on the other side of the creek, where the bank was three feet high and the water ran two or three feet deep. Dark as it was, we could see the water splashing and hear something trying to get up the bank, but could not make out distinctly what it was, although from the snort we had heard we knew that it must be a bear.

My companion was packing three or four salmon that we had caught and were going to put out for bait, while I had an eight by ten camera on my back. There was a sudden dropping of fish and an attempt to shed the camera, but by the time this was done the bear had climbed the bank. For an instant, however, he appeared silhouetted against the western light, and I saw his gray coat and took a quick shot at him. This brought out another snort, with much breaking of branches, and we could hear him for quite a while as he charged through the brush in the bottom.

There was no use in trying to follow him in the dark, nor, indeed, would it have been advisable, and we went on to camp, thinking to trail him in the morning in case we found that he was hit. But in the night it began to rain, and it was still raining when we started out in the morning. Under some thick trees, where the rain had not washed it out, we found plenty of blood, and at other places the bear had lain down for some time, but after we got out of the thick timber rain had washed the blood off the grass and leaves, so that it was impossible to follow the trail. We cross-cut the bottom and skirted the base of the mountain, but we could not find the bear, and finally we gave up the search. Then, some three days later, our cook saw several buzzards circling about and swooping down over near the mountain, and he went out to see what it all meant. He found, between two large boulders that were nearly overgrown with brush, the body of the old grizzly that we had wounded. He had crawled in between the rocks and had covered the entrance so completely that, though we passed it twice at least, we never saw it. The hide was quite useless. We took the teeth and nails of the poor old fellow, but I would gladly have returned them to him with his life, if I could have.

IX

TRAILING

TRAILING the grizzly bear is, in the language of the regions which he inhabits, "no cinch." Of all the forms of still hunting or stalking open to the sportsman of America, it is the most difficult and the most demanding. Not only is the grizzly phenomenally quick to catch every sound, not only is his sense of smell amazingly developed, but he is particularly cunning in guarding himself against danger from the rear, and his senses are at least matched by his shrewdness.

In following the trail of any wild animal, an observant person soon begins to learn the peculiarities of his quarry. Especially if there be a slight coating of snow on the ground the record of recent hours is plainly written, for him who trails to read. And as "Moccasin Joe" is at once more human in his proceedings and more capricious in his occupations than any of his fellow denizens of the forest, as he exceeds these also in cunning and endurance, the tracking of an old grizzly, made wise and wary by years of experience, is the most searching test of a hunter's skill, and offers him, both in entertainment and satisfaction, the greatest reward.

Upon finding the trail of a bear, if the knowing hunter

77

will take his time and go along leisurely and without making too much noise, he will, if the bear has not too great a start of him, be reasonably likely to obtain a shot. But he will find that cunning has to be matched with cunning, and that unless he keeps his wits on watch and remembers that he is not tracking some stupid beast that takes no thought of who or what may be behind, the bear will quietly make a détour, sniff the wind that reveals the presence of the huntsman, and then, abandoning less pressing interests for the time being, turn his attention to preserving his own hide from capture, and either hurriedly cross the divide or take to the thickest jungles, to lie low until the coast is clear.

Once the hunter is discovered it is usually just as well to look for another bear. I have always found a chase under such conditions to be a long and useless one, and not likely to result in the death of a bear once in a hundred times. Twice only have I bagged my bear under such circumstances; and then it was in a country where I was able to cut across and head him off as he passed along the side of a mountain.

The hunter's first concern, of course, is to determine as closely as possible how long it has been since the bear passed by. Then, unless the wind is in the right direction for direct trailing, wide détours must be made to the right or left, striking the trail far ahead. If the bear is still going in the same direction, another détour must be made, and this must be kept up (sometimes for several days) until the bear is either sighted or has taken a course that the trailer can follow without fear of the wind's carrying his scent to the game. This method of roundabout

From a photograph, copyright, 1909, by J. B. Kerfoot

LISTENING

trailing is not as interesting as it is to follow the trail direct, but it is much safer and more likely to crown the hunter's efforts with success.

In following one of these trails one will find, every little while, where the bear has made a circuit to the side and rear, so as to get to windward of his own back trail, and so assure himself that no danger follows on his track.

The bear is wary and he often will make a détour of a mile or more, and will stand quietly for some time until satisfied that there is nothing of a suspicious nature within reasonable distance of him. Then, if it is autumn and in the mountains, he will resume his hunt for gophers and marmots.

The grizzly usually goes into winter quarters among the rocks, at a much higher altitude than does the black bear, and as hibernating rodents are the only food he can find along the high peaks and ridges, he devotes much time, at this season of the year, to digging them out and devouring them. Throughout the Bitter Root Mountains, and in the Selkirks of British Columbia, the last food that the grizzly obtains before denning up are the hoary or whistling marmots, and the little animals commonly called gophers, but which, properly speaking, are Columbian ground squirrels. These latter inhabit not only all parts of these mountains, from the highest and most rocky peaks to the lowest valleys, but are also a great pest to the farmer. One of them is about as large as an Eastern gray squirrel, but they are slightly different in shape, being less long and slim, and in fact more like woodchucks —short-legged and very plump and heavy of body. The hoary or whistling marmot is much larger than the ground squirrel. Many of them grow to weigh twenty or twenty-five pounds. They are about the size of a very large badger, and as several of them often live in the same den, the bear, if he can reach them, obtains quite a feast. It always seemed to me, however, that in the case of the ground squirrel, the grizzly's game was hardly worth the candle; for at best one of these little rodents is but a mouthful for a bear, and in many instances I have found where, after digging out cartloads of dirt and rocks, the bears had had only their trouble for their pains. But no amount of labor seems to daunt them. I have seen many such holes that were from eight to ten feet deep and twelve or fifteen feet long, where one or more grizzlies had thus dug for a nest of marmots. And a few years ago, while hunting through the Selkirks with Mr. G. O. Shields, we came across a tremendous hole thus made by grizzlies. There were literally carloads of dirt and rocks taken from the opening and piled up on the mountainside, but whether success had attended the diggers' efforts, and the marmots had been captured, we were unable to determine. When snow is on the ground it is usually possible to tell whether a capture has been made from the telltale drops of blood.

It is at this season that I have most enjoyed trailing the grizzly, not only because his tracks are plain, but because so many of his doings are written by the way.

Some years ago, while hunting along one of these high divides, I came across the track of a large grizzly that was hunting squirrels and marmots. As it was quite early in the morning I decided to put in the day, if necessary, trying to get him. I saw that the bear had passed along either during the evening before or some time in the night. I knew this because the snow that had been displaced by him had frozen later and the tracks had not thawed out any, as (the day before having been warm) they would have done had they been made then. I judged the bear had from six to twelve hours the start of me. But this did not discourage me, for I knew that he would spend many of them in digging for food, if he found a promising lead.

Near the point where I found his trail he had dug for a gopher, but after getting down some two feet he had struck a rock and had given it up. From there he had travelled along the side of the mountain, stopping to examine all the small bunches of bushes and loose rocks for marmots. Here and there he had turned over a rock, but as there was little likelihood of finding any grubs or worms at this time of the year he only stopped to explore the dens of the gophers. For several hours he had moved along slowly, as I could tell from the fact that his steps were rather short and that every little while he had stopped and stood still, presumably for several minutes.

Twice he had made a détour up the side of the mountain and had then trailed back for half a mile to a point where he could sniff the wind and obtain a view of the country he had passed through. Then, on the latter of these occasions, he had made himself a bed in some bushes and had slept for several hours, as shown by the snow that had been melted from the heat of his body.

After his nap the bear returned by the trail he had made, walking along back to where he had turned off to make the détour. This they sometimes do, but not often, and when I first saw this track I thought that another bear had come into the game, but found, on following it around, that it was the same bear.

From this point the animal went in rather a straight line along the mountain for several miles. Then he found the den of a family of marmots which he proceeded to unearth. For half a mile before reaching the place I could see the dirt piled up on the snow, and knew I had gained several hours on the bear. The hole he had dug was larger than miners are annually required to excavate in order to hold a mining claim. The den ran in under several layers of loose flat rocks, some of which were two or three feet long by half as many wide, and several inches thick. These he had ripped out easily and thrown down hill, and the dirt and small boulders had been hurled out and now covered the snow all about for a space of ten or twelve feet.

On the rocks and snow were large spots and blotches of blood, telling of the feast that had rewarded his labors, and that there had been more than one marmot was shown by the numerous tracks. These animals had burrowed down some six or seven feet into the side of the mountain, and under a large flat stone they had scooped out a little cave, some three feet in diameter, where they had a soft bed of grasses that they had carried in. When the grizzly broke his way into their home, there had been a great rush for freedom. And as the sides of the hole dug by the bear were rather steep, the marmots, in trying to escape, were at a disadvantage. The whole story was plainly written in the soft earth and the snow. Large impressions in the sides and bottom of the hole showed where the bear had struck at the flying marmots. One of these latter had evidently succeeded in making the outside, but the tracks of the bear showed where he had followed in hot pursuit, and a smear of blood on the snow marked the end of the little fellow's dash for life. After killing and eating this one, the bear had returned to the marmot den and devoured the remaining marmots already killed; after which he had come out, walked around for a while, and then struck out again in search of more provender.

For a mile or more he had now kept along the same ridge, then turned to the right, crossed the divide, and made off down the side of the mountain, without stopping until he had crossed the bottom and was well up the opposite slope. On this ridge he had stopped and taken another nap, and I knew that he was now but a short distance ahead of me, for, although it was well along in the afternoon, and the snow had become so soft that it would pack, it had not melted any in the bear's trail since he had left his resting-place.

The question was, could I catch up with him before night? To do it would require all the caution and skill I possessed, for it was evident from his movements that I was dealing with a bear of the "old school."

The animal now kept along the ridge for several miles, going back in the direction from which he had come the night before. There was apparently no wind, but I was taking no chances on this score, and whenever I came to a patch of bushes or timber, I made a détour up the mountain to a point where I could see well below me, and then circled until I had found where the bear had passed.

I would then search carefully the open hillside as far as possible before moving ahead on the trail. Once I was able, with the aid of my field glasses, to make out his course for nearly a half mile, where he had climbed up the side of an open hill.

By this time, however, the sun had gone down and trailing was now not only necessarily slow work, but bad policy, since a wrong move would spoil my whole game. I was ten miles or more from camp, and only a short distance behind the bear, yet I did not dare follow him, for fear he would discover me and I would lose him altogether. I therefore turned down the mountain into the bottom where there was a little creek, built a small lean-to of fir boughs back some ten or twelve feet from a large boulder, got up plenty of wood, built a fire against the rock, and ate the lunch I had brought with me. The lean-to sheltered me from the wind, the heat reflected from the rock made my camp comfortable, and when my fire needed replenishing during the night, Jack Frost woke me up to attend to it.

At the first streak of day I started up the mountain to the point where I had left the trail the evening before, and took up the stern chase with renewed interest. I followed up mountains and down gulches for miles without finding where the animal had stopped to dig for anything, and I began to think that he had made up his mind to leave the country. At last, however, I saw a pile of dirt far ahead, and knew that he had stopped at least for a short time. The hole was not so large or so deep as the first one, but it had evidently yielded a mouthful or two of ground squirrel for breakfast. Farther along he had dug other small holes for squirrels, but none had detained him very long.

But on the next ridge I found another bed, and knew, since the snow was not frozen around it and there was no frost on the leaves and grass where he had lain, that bruin had left it after daybreak. Then, too, the heat from his body had melted the snow and left the ground damp, and as this had not had time to freeze I knew that he had but recently moved on. In the next mile he had only stopped once, and once he had made the usual détour to see if all was well, and had then wandered aimlessly about for a while and finally turned toward a thick clump of bushes and stunted trees that grew on the side of the mountain in a small ravine.

Searching the side of the hill with the glasses, I could plainly see where he had entered the brush, but could not see where he had left it. There was, however, one point of possible exit that I could not see, so I climbed up and around the spot, and looking down from above, found that the bear must indeed be there, and within shooting distance. Selecting then a position from which I commanded all the strategic points, I rolled a small rock down the mountain nearly, but not quite, into the bushes. This went bounding down, making much noise, and soon I saw the small trees near the centre of the brush move, and knew that the bear was on the alert. I could follow his movements by those of the bushes, and when he showed his head I was ready for him. He did not come out with a rush, but walked quietly to his death; and as his head pushed through the edge of the bushes, one of my old six-hundred-grain bullets met him at the butt of the ear, and the trailing was over. He never knew just what it was that had awakened him, nor did he ever know just what it was that put him back to sleep.

X

A CHARGING GRIZZLY

FOR several years I had hunted grizzlies by stalking, and by watching their feeding grounds, and in this way killed many, but many also got away. Sometimes I would see as many as five or six in a berry patch and bag only one or two; sometimes I got three; once I got four; and once, as I have shown, I was able to get five. Yet on one occasion I saw nine large ones together, and succeeded in killing only one, and that the smallest of the lot. So that I thought it might be a good plan to get some bear-dogs, and accordingly set about finding out what I could about them.

In one of the sporting magazines about this time I saw bear-dogs advertised for sale, and a letter to their owner developed a correspondence that finally resulted in a hunting trip with a customer to whom he had sold a dog. This customer, Mr. John D. O'Brien, in due time arrived at Spokane with the bear-dog and another hound that he had brought along to be taught bear hunting. The bear-dog was a mongrel, but of what breeds no living man could guess. He answered to the name of Nebo. The third, human, member of the party was Mr. Martin Spencer, and I secured from an old Indian two likely looking pups, one of which he claimed was a bear-dog. It was a regular Indian dog of about five or six colors, and proved one of the best bear fighters I have ever seen. The other was a cross between a brindle bull and a staghound, and, while he had never hunted any, he could whip anything, aside from a bear, that he came across.

For a hundred and fifty miles our way lay across the Palouse farming country. We then crossed the main fork of the North Clearwater River and struck into the Bitter Root range from the west. Nothing of interest occurred until we struck the hills, when Jack's hound jumped a deer, and I suppose he is still after it, as he never returned.

We soon, now, began to see bear tracks, and in order to give the Eastern bear-dog all the chances there were, Jack thought it better not to let my curs loose until his dog had brought the bears to bay. When we showed Nebo a bear track he would grow vastly excited, disappear up the trail, and after a few minutes we would hear him barking. But when we reached him we always found that he had treed, not bears, but "fool hens"—a species of grouse. And for two months we followed that dog uphill and down, and not a bear did he tree. However, we killed a number of black bears without his assistance.

One evening after a hard day's climb over mountains where there was no trail, we camped on the west side of a bottom, some hundreds of yards from the large stream that ran through it. The bottom contained several hundred acres covered with high grass and willow bushes, and as we got into camp early I took my gun, after unpacking, and went up the stream to hunt for a way out, since we were cross-cutting the country without regard to trails. I soon came to a small stream, blocked in several places by beaver dams, and in part of the bottom, which was flooded nearly a foot deep, I noticed where some animal had a wallow. At first I paid no attention to this, thinking it was a moose wallow, but after crossing the creek I saw the tracks of a large grizzly and of several smaller ones, and in a clump of willows farther on I found several beds where the bears had been lying. However, I saw nothing of the bears and, having got the lay of the land, I returned to camp.

Upon my arrival I told Jack that there were some grizzlies out in the willows, and suggested (this having become a standing joke) that his dog might be able to find them. I also told him that there were some beaver in the small creek, since he had been abusing the country for its lack of beaver, and had been anxious for some beaver-tail soup, having heard that it was a great luxury. Taking his gun and dog, Jack now went over to the willows, thrashed around for a short time, and returned without seeing anything. He said that the tracks were old, as Nebo would pay no attention to them, and tying up the dog, he went back to the creek to watch for beaver. After a while we heard a shot, and soon he returned to camp, telling Martin that he had wounded a beaver, but that it had escaped to its house in the stream, and he was unable to get it. Martin proposed that they go and tear the house to pieces and thus obtain the animal, and Jack again loosed the dog, shouldered his gun, and, Martin taking a long-handled shovel which we had brought along to prospect with, they set out.

Jack's gun was a single-shot, .40-70 Winchester. He claimed that a good hunter should never carry more than four cartridges for a day's hunt, and consequently he never carried but that number. One of these was always in the gun, the other three were tucked away in his pocket, one in each finger of an old kid glove.

That evening there were three old miners camped near us, and in about ten minutes after Jack and Martin left, we heard the dog bark and the report of Jack's rifle. In a short time there was another report and the barking of the dog continued. Almost immediately another shot was heard and the miners began to wonder as to the cause of the shooting. I told them that the boys were after beaver and that, as Jack never took more than four cartridges with him, the fun was about over. And just then we heard the report of the last shot. I was about ready to turn into bed, as I was tired after the long day, and had already taken off my shoes, when I heard a shout and saw Martin coming through the flooded bottom ten feet at a jump, splashing water as he came. His hat was missing and he acted like a limited express going through without stops.

And as he came he yelled: "Get your gun! Turn the dogs loose! There are some grizzlies out there and I guess they've killed Jack," I managed to make out. So I loosed the dogs, caught up my gun, and, without waiting to put on my shoes, made the water fly nearly as high as Martin had in coming in. I jumped the small stream, and, landing among the stumps of some shrub willows that had been eaten off by the beaver, punched several holes in the bottom of my feet. I could not, however, stop for that, and as I hurried on I could hear the water splash behind me as Martin, having secured his rifle, hastened after me. Then, as I came into the open, I saw Jack standing at the edge of some timber, leaning on his rifle, while in front of him, sixty feet or so away, a large grizzly lay rolling and bawling as she rolled. In the willows across the creek there was a big commotion—bears bawling and Nebo barking; my curs joined in the uproar; a bear showed his head and I sent a slug into it; Martin came up in time to get a shot at another; a third appeared and was instantly killed. We had four bears down and there seemed to be no more coming.

Martin, in telling of the beginning of the adventure, said that when they came to the small creek and looked over the beaver house, they decided that they would have to cut away a dam below it and run off the water. Jack, therefore, placed his gun against a tree and went forty or fifty yards down-stream to the dam. As they were about to begin cutting this away, they were attracted by a noise in the willows across the creek, and Martin had advised Jack to get his gun, thinking it might be a moose. However, as they heard no further noise, they went to work on the dam, and when they again looked up they saw an enormous grizzly standing on its hind legs and looking over the willows at them. The dog had then begun to bark, and the bear, accepting the challenge, started for them. Jack sprinted for his gun, and Martin, sticking his spade into the mud, shinned up a tree, prepared to referee the show. As Jack got his gun the grizzly appeared at the water's edge directly opposite, and before he could shoot, jumped for him, landing in the middle of the stream. Jack now put in a shot, and as his small supply of ammunition was unhandily packed, was compelled to strike out for the timber, digging for cartridges as he ran.

The bear-dog had been very bold when the bear was on the opposite side of the creek, but when it forded the stream Nebo with great discretion retired some two hundred yards across the bottom; and had Jack's life depended on the dog's attacks, he would probably have been slaughtered then and there. And yet, to do Nebo justice, his barking saved Jack's life in the end. Martin, from his perch, could see Jack's effort to reach the timber, and to reload his gun before the bear overtook him, and, as the latter got nearer and nearer, he called, "Look out, Jack, she's after you!" Jack, with this stimulus, made the timber, and the barking of the dog seems to have turned back the bear. Jack now got at his cartridges and, coming out from behind his tree, took another shot at the grizzly from a distance of about thirty feet, but missed her, and she, turning, again went after him. Martin, from his observatory, again called out the news, and again, after Jack had reached his shelter, the barking of the dog drew the bear off. Jack reloaded again, fired again, and missed again, and this time the bear went for him in earnest.

Martin still acted as a bureau of information, but as the bear was now so close to Jack that he had all he could do to dodge her claws, this was hardly necessary. This time he had kept his cartridge out, and he loaded as he ran, and reaching a large tree he swung himself around it, and as the bear passed, fired his last shot—and made a clean miss. He was now forced to continue his flight, and as soon as he was out of sight Martin made for camp.

But Jack never got scratched. After the bear passed the tree she went back to the creek, having now begun to feel the effects of the first shot, which, entering her side, back of the ribs, had ranged downward and passed out of the groin between the hind legs. That it must have been the first shot that struck her is proved by the downward course of the bullet, fired from the bank when the bear was in the creek.

This is the only instance that I have ever known personally where a grizzly attacked a man without provocation, and in this case I lay it to the dog. The bear was an old one with three yearling cubs, and the presence and barking of the dog were naturally offensive to her. And as the cubs and the dog were doubtless the reasons for her unprovoked attack upon the man, so also the cubs and the dog proved his salvation. For the bear left him again and again and returned to interpose herself between the barking dog and her cubs, which had not crossed the creek.

In this case, too, the vitality of an old and large grizzly proved to be many times less than that of the deer. Not a bone was touched by this first bullet, and not another ball hit her. Of this we were sure, as when we came to skin her we examined her very closely. Jack was fully convinced that he had hit her three, if not four, times; and had she not been killed, he would have sworn that he had so hit her, and that, in spite of her wounds, she had made good her escape. Twice he had fired at her at a point-blank range of thirty feet, and his last shot was delivered when she was but ten feet away. This experience has, in my mind, raised the question as to how much of the lead a grizzly is commonly supposed to carry off has ever really reached its mark.

I may add that I heard, long after this, that Nebo finally (by accident, I presume) came across an old bear blocking a trail, and absent-mindedly trying to remove the obstruction, achieved a bear-dog's finish.

XI

AT CLOSE QUARTERS

LATE in the fall of 1891 I took Dr. C. S. Penfield and James H. Adams of Spokane into the Bitter Root region after big game. I had just returned from a bear hunt in which our party had killed thirteen grizzlies, and as neither of these gentlemen had at that time killed a bear, this made them particularly anxious to get some. We took my two bear-dogs, the bull, and the Indian mongrel with us, and entering the mountains from the west, we followed the old Lolo Trail that keeps to the crest of the main ridge from four to five thousand feet above the valleys on either side. It was late in the season, and, the bear having sought the lower altitudes, we did not find any along the trail, nor, though we camped beside it for a couple of days and hunted elk and moose, did we have any success. We therefore followed the Lolo for two days more, meeting several parties of Indians returning, heavily loaded with elk and bear meat, from their annual hunt, and then we turned to the right, descended some five thousand feet, and camped on the banks of the middle fork of the Clearwater River.

This stream is about one hundred yards wide, is, at this season, quite shallow, and like all the streams in this country, very rapid and very cold. Its banks and bed are lined with large boulders, and the water makes a tremendous noise in running over them. Our camp was on a narrow strip of flat land that paralleled the river for half a mile, and ran back two or three hundred yards to the foot of the mountain. Just above us the Clearwater made a turn almost at right angles, and in the bend, on a little patch of open ground, with two large trees in the centre, the Indians for years had made their camp, and the place was marked by piles of refuse and by their racks or scaffolds for drying meat.

After unpacking our horses, I sauntered up to this opening to see if the Indians had left any elk antlers—they hardly ever carry them away—and the bulldog trotted after me. As I emerged from the bush I saw, behind a bunch of bushes and near the drying racks, what at first I took to be a black stump. Remembering, however, that I had camped at this place many times, and had never seen a stump there, I took a closer look and saw that it was an old bear gnawing the discarded bones. The animal was so dark in color that I supposed it was a black bear, and determined to tree it for the doctor or Adams to shoot. I had no weapon of any sort with me, as I had taken off my belt in unpacking the horses and had thrown it, with my hatchet, on the ground, and had left my rifle standing against a tree. I had, however, no idea of killing the bear, and even had I had my gun, would not have robbed the others of their opportunity.

The bear was near the two trees, and I thought there would be no trouble in putting him up one of them, so calling to Jim, the dog, to "sic 'em," I started toward him.

ELK SUMMIT, BITTER ROOT MOUNTAINS

But the dog had not yet seen the bear and, dog fashion, simply ran back and forth looking for he knew not what, so that I was soon within fifty feet of the bear and, not caring to quite jump on him, I gave a yell; whereupon he stood up to see what had broken loose, and Jim got sight of him and made for his hind legs.

I now began calling at the top of my voice for the doctor and Adams, while the bear moved off toward the trees. But as he had to pay attention to the assaults of the dog, he could not run very fast, and I easily kept up with the procession, yelling the while for re-enforcements. But the bear did not seem to have the least intention of climbing. He stopped under one of the trees and made a few passes at the dog (I, meanwhile, getting in a few more yells), and then he moved on toward the river, the bank of which was some thirty feet above the water, and while not straight down, was so steep that it was quite difficult to negotiate either way.

We came out about two hundred yards above the bend and just where, along the bottom of the bank, there grew a few old cottonwood trees. Here Jim grabbed the bear from behind, and the latter, in turning to strike him, lost his footing, and bear and dog both disappeared down the bank. I ran to the edge and seeing them bring up against the old cottonwoods, thought the bear, who must by now have had enough of fighting, would surely climb. But instead he plunged into the river, and as soon as he got to where the water was deep, turned and faced the dog. This, as the bear seemed to know, put Jim, who had to swim, at a disadvantage. But the current was swift, and swept them both down-stream to where, at the head of a riffle, several large boulders stood out of the water. Jim crawled out on one of these and the bear sat down just above it, and every time the bear started to move Jim made a rush for the edge of rock and the bear turned back.

On the opposite side of the river there was some heavy timber, and I thought that if the bear got into this he would surely tree. So I determined to drive him across, as, once in a tree, I could watch him until the men at camp hunted me up. I figured that if I did not return before night they would start a search, and if my voice did not give out, they would soon find me.

In fording a swift stream where there is danger of being swept off one's feet, a stout pole, used as a brace downstream, will often make easy a passage that would be impossible without such aid; and I now, therefore, looked about for a pole, and finding one, started to drive the bear across the river. When he saw me coming he started for the other shore, and Jim, leaving his boulder, followed close behind him. My own crossing was a more difficult matter. The water was swift, the boulders were slippery as glass, and it was all that I could do to make headway. By the time I got across, the bear and the dog had both disappeared in the brush, but an occasional bawl told me that Jim was still on the job; and hearing a bark behind me, and seeing old Pete, the mongrel, just entering the water, I felt sure the hunters were coming and started contentedly ahead.

I soon came up with Jim and the bear, and whenever I saw an opportunity to do so without getting cuffed, brought my pole down on Bruin's back. The brush, however, was so thick that it was hard to do the bear any harm, or indeed to produce any effect beyond a snort and a more than usually vicious stroke at the dog; and my pole soon became so shortened by constant breaking that only a small club remained.

We now came to a little stream that drained a spring fifty yards back from the river. Across it there was a fallen tree about eighteen inches thick, and across this tree, and parallel with the creek bank, another tree had fallen, forming a sort of pen ten or twelve feet square. Pete now came up, the bear and the dog stumbled against these logs and fell into the pen, and as they tumbled over I happened to notice the bear's paw, and for the first time, and to my utter astonishment, realized that it was a grizzly, instead of a black bear we were trying to tree.

I now saw that I was probably in a scrape if those hunters did not show up, and I tried to yell louder than ever, but I had done so much of that already that my voice was nearly gone. However, I stood outside the log pen and did the best I could, and now and then, when I thought the bear was not looking, I whacked him over the head with my club, but of course did him no harm. Jim, the bulldog, was on the side next the creek, while Pete, the mongrel, was over next to the log. When the bear came Pete's way he slid under the tree out of reach. Jim, on the other hand, was not so lucky, as he had to jump the two-foot bank, and I cannot to this day see why the grizzly did not kill him.

I saw that if I had a good two-handed, heavy club that would not break I could likely smash the bear's head, and regretted taking off my belt, on which a small hatchet was slung. I never carried a hunting knife, but did all my skinning with a pocket knife containing one three-inch blade and one smaller one, and as this was the only resource at hand I got it out. I was looking around for the sort of club I needed when I heard a snort, and looking back saw that the bear had decided to turn his attention to me. I sprang back, caught my foot in a root, and fell flat on my back in the creek, with my head about a foot lower than my heels. "Now," thought I, "you are in for a good chewing," and being unable to take any other precaution, I grasped a bush in my left hand, got a good grip on the knife, and determined to run it into the bear's belly and open him up. But when the bear got his front paws over the log, with his nose just at my feet, both dogs grabbed him by the flanks, and bear and dogs were all tumbled back into the pen.

As they fell I made a thrust and drove the knife in behind the bear's right foreleg, and the blade slipped in so easily that it gave me an idea. I got up and, placing one knee against the log and the other foot against the root that had tripped me, I waited my chance and made another thrust. Had I had a long-bladed hunting knife the fight would have ended then and there, but the knife was short and the bear fat. So that the stab was only an incident.

Up to this time I had, in a desultory sort of way, kept on yelling; the bear, when he got an unusually hard nip, had given an occasional bawl; and now one dog and now the other had given a yelp when he feared the bear was about to get him. But now these sounds were hushed. Each of us had his work cut out and we got right down to business. I kept my place by the log, and when the bear turned to attack Jim, it was my turn to stab. When the bear went for Pete I let him most severely alone, but this was Jim's opportunity and he never missed it. After the bear had been stuck by the knife a number of times he became foxy. He would feint for Jim, and as I started to make my thrust, would turn like lightning and make a pass at me. Several times he came so near me that I thought I could hear the swish of his nails, and I was soon compelled to be more cautious and to wait until Jim became actually entangled with the brute, when Pete and I would join forces and make the most of our opportunity, till we drew the fight back to our quarter.

Jim was now becoming winded; he could no longer retreat as quickly as at first, and I myself was beginning to miss the breath I had spent in yelling. Pete was the only one of the attacking force that was in first-class shape, while the bear seemed to have more wind than he knew what to do with. Under these circumstances it was not long before he made a pass at me and caught me through the ball of the hand. The wound was rather a nasty one and the blow knocked the knife from my grasp; but after striking a bush it dropped near by, so that I was able to recover it without losing my next turn, and we continued the fight. Little by little the bear himself weakened. At first this was only noticeable in that it required more biting and punching than formerly to make him turn from one to the other. Then bloody froth began flying from his mouth. But he held on so amazingly that I about lost hope of killing him and saving the dogs. Still I, had got them into the scrape, I knew that they would fight to the death, and I made up my mind to stay with them.

Finally I saw a change come over the bear—the change that marks the transition from doggedness to desperation —and I knew that it would likely go hard with the next one he grasped. This proved to be Jim, and wrapping his forelegs round him the bear dropped on his side and began trying to rip the dog up with his hind claws. Pete dove to the attack from under his log, but the bear paid scant attention, and, hearing Jim's smothered cries, and the bear lying with his head and back toward me, I grasped the knife handle by the extreme end, laid my left hand on the bear's head, and in a last, desperate effort to reach a vital spot, drove blade and haft both into the bear's side. Then I leaped away, and I was not a second too soon. The stroke—as though it had released a spring—brought the bear convulsively to his feet, and I barely missed the mighty stroke he aimed at me. But this time my thrust had gone home, and, the force of his own blow carrying him off his feet, the grizzly fell prone across the logs and the fight was over.

I had received no hurts other than the one in the hand, and although I was covered with blood, little of it was my own. Pete escaped without a scratch, but Jim, as far as appearances went, was worse off than the bear. One of his legs was so chewed up that he could not use it; his neck was lacerated until it seemed as if his head was cut half off, and there were several other severe cuts about his body; but he was not subdued by a good deal.

For a few moments, bloody and panting, I sat on the fallen tree and watched the dogs worry the dead bear. Then I rolled the carcass over and out of the arena and, the way being a little downhill, and finding that by reaching back and grasping a side of the head with each hand I could just drag the body along, I started with it for the river. Several times in the short distance I was brought to a sudden stop, and on looking around found Jim, the bear's hind paw in his mouth and his own three legs braced, pulling back for all he was worth. Arrived at the bank, I rolled the body in and succeeded in towing it across, but being unable to lift it out of the water, I drew it in behind a ledge of rock, left it there, and went on to camp.

There, very much at their ease, sat the two hunters who were so anxious to kill a bear. When they saw me, wet and covered with blood, they became excusably excited and wanted to know what was the matter. I told them that I had been killing a grizzly. "But what did you kill him with?" they said. "Your gun is here in camp." "Well," I answered, "as you fellows did not come when I yelled for help, I had to kill him with my pocket knife." But it took the body of the bear and a post-mortem to boot to satisfy their doubts.

It seemed that, as soon as I had left camp, they had gone down to the river to get a drink, and the noise made by the stream prevented them from hearing me. Once, they said, they thought they heard a dog barking, but as the sound was not repeated, they thought no more about it. When we came to skin the bear—he was a handsome animal, his pelage a deep black, scantily touched with white—we found seventeen knife wounds back of the right shoulder. Three of his ribs were completely severed and the last stab had pierced his heart. Jim and myself were considerably knocked up. The doctor sewed up our wounds. Jim had to make the rest of the trip on three legs, while my right hand was out of commission for a while and I had to talk in whispers. And as it turned out, this was the only bear we saw.

The next day it began to snow and rain, and kept it up for a week, and when we climbed the five thousand feet back to the trail, we found the snow so deep that we came near having to abandon the horses, and of the six we had with us got only three home.

XII

MY FIRST TRIP TO THE SELKIRKS

FOR years before I visited the district I had heard wonderful stories of the grizzlies of the Selkirks. I had heard how plentiful they were, and how ferocious they were, and how many miners were killed by them every year. I had met old prospectors and trappers who insisted that several species were found among them—the Roach Back, the Silver Tip, the Bald Face, and now and again a Range bear. Of these they agreed that the Bald Face was the most dangerous and the most aggressive, and declared that he would turn from the trail for neither man nor beast, but, with a chip on his large shoulder, was always on the lookout for some prospector whom he might mangle for the fun of it.

I confess that these reports attracted me. I thought that the chance to hunt bears where the sport did not require so much sneaking and crawling as I had been accustomed to would be much to my liking; and I determined to take the first opportunity to visit a region where the bears really themselves went hunting for the hunter. Accordingly one spring, in company with Mr. T. C. Coleman of West Virginia, and Dr. C. S. Penfield of Spokane, both old hunting companions, I made a trip to the Wilson's

Creek country, a part of the Selkirk range in British Columbia.

We left Spokane about the middle of April, but upon reaching the hunting grounds found that we were about three weeks ahead of time, since in these northern hills the bears do not come out of winter quarters so early, by nearly a month, as in the more southern ranges where we had hunted them for many years. Indeed, we found that conditions in the Selkirks differed altogether from those of the country in the south. The lay of the land was different. The food of the bears was different. And the methods of hunting, which depend largely upon these two features, were necessarily different in proportion. The only factor that remained comparatively constant was the nature of "Old Ephraim" himself.

In the Selkirks the mountains are very steep, and their sides come down in the form of a sharp V to the bottoms of the cañons that do duty for valleys between them. In these rocky gorges run very rapid streams; and down their steeply sloping sides in the spring, the snows rush in great avalanches. These move with terrific force, sweeping before them anything that offers resistance, and are apt to follow the same tracks, year after year. And it is in these tracks, locally known as "slides," that, from necessity, if not from choice, one hunts for bear.

The grizzlies in this region very seldom get any animal food, but live almost entirely on roots, bulbs, grass, and the buds of the small maple bushes. In the early spring, when the grizzlies first come out, the snow is piled up in astonishing fashion at the bottom of these slides, and below, and on either side of them, covers the country to a depth of

NEAR THE HEAD-WATERS OF WILSON'S CREEK

from six to ten feet; but the slides themselves are nearly free from snow, and here the bears are forced to come to look for food.

The slides are thickly grown with many kinds of brush, and here and there show little parks where, on the southern slopes, the grass sprouts early. Here, too, grow two plants, the dog-tooth violet and the spring beauty, each springing from a little bulb like a small onion, and the ground is sometimes so dug over and torn up by bears searching for these bulbs, that it looks as though a band of prospectors had been at work.

Strangely enough, though both of these plants grow in the Bitter Root Mountains, I have never seen signs of a grizzly's having dug for them there.   On the other hand, a plant called the shooting star, with a leaf like a small horseradish leaf, which also grows plentifully in both places, is greedily eaten by the Idaho grizzlies, and wholly ignored by those of the Selkirks.

The usual way of hunting in these Selkirk valleys is to make one's way along one side of the streams and scan the slides across the gorge.   The hills are so steep and the brush so thick that it is impossible to see anything on one's own side of the creek, and indeed it is hard enough when one does see anything, to make one's way to within range. For the brush, what with the weight of constant snows, and repeated bendings of the head to passing avalanches, all grows downhill, and it is all but impossible to worm one's way up against it.   We cut many a mile of trail and felled many a foot log across the streams, before we got through with Wilson's Creek.

However, when we first reached this forbidding and

inhospitable region, not only had the grizzlies not yet put in an appearance, but we ourselves were stalled ten miles below our intended destination. We had already travelled some miles through snow from two to four feet deep, and now, being unable to make further headway, we cut down enough timber to open a clearing for our tent, unpacked the horses, and sent them, with the man who drove them, back to the settlements where we had hired them.

And now the weather, which had been warm and pleasant, turned suddenly bitter, and for two weeks we three held our little camp against a wintry siege, and when this finally was raised and pleasant days were promised, the doctor's leave of absence had expired. He therefore walked out to the settlements, ordered our horses sent back to us, and left for home, and the next morning, our pack animals having arrived, we moved on up-stream and made our intended camp.

On our way up we saw our first bear track of the season, the animal having evidently passed only the evening before, and worked along the edge of a large slide, where here and there there was a patch of bare ground. The next morning, when the packer had taken the back trail for home with his horses, Coleman and myself took up the old bear's trail and tried to follow it. After crossing the slide, the bear had stopped several times to dig up small roots and bulbs, but as there was little vegetation to speak of, he had worked around toward the south side of the mountain, where there were several other slides. It was very difficult to trail him where there was no snow, and we soon decided to drop the trail and strike straight for the southern slides, thinking to find him on one or the other of them.

After several hours' slipping and struggling, and repeated disappointments at several likely looking slides, I crept across a welcome bed of soft pine needles to the edge of a big opening, and craning my head carefully forward, looked about me. Now, it is one thing to creep forward yourself to get a shot at a bear, and quite a different one to take another man to where he can obtain the shot. Coleman always kept in the rear, because, he said, he was not quick enough at sighting game; so, when we started out, it had been understood that he was to keep close to me, and in case there was any chance of a shot, I would reach back, seize him, and place him in front.

When I poked my head through the brush on the edge of this slide, the first thing I saw was our old grizzly only fifty or sixty feet away. I therefore reached silently back for Coleman and, not touching him, turned to see what had become of him, and there he was some forty yards in the rear. I made frantic motions for him to come on, and seeing from my actions that something was in sight, he hastened up as rapidly as possible. As soon as he reached me and had regained his wind we looked out over the slide, and were dumfounded to see it tenantless; the bear had gone. We went out and found the trail, followed it into some cedars fifty yards above us, and found that it then made up-stream the way we had come, and that, had we remained at a little creek where we had rested on our way down, we would have got a shot, as the bear had passed close to where we had been sitting. We trailed him for about a half mile, and finding that he was taking advantage of all the thickets he could find, and that he was evidently on the jump, we decided that it was more than useless to follow. So we returned to camp.

The next morning we started up-stream on snow-shoes. When we had gone about five hundred yards I looked up at a slide on our left, and there was a huge grizzly walking over a mass of snow that had slid down about a week before. We watched him, and as he soon began feeding on the grass that had started to come up along the edge of the snow we determined to make a try for him. We found a fallen tree across the stream and, clambering over, shinned up the steep bank and then crawled through the bushes and came out at the spot where we had seen him feeding. But no bear awaited us. He had evidently got wind of us in some way and moved quietly away.

Since now the snow was too deep and soft for us to get through the brush without snow-shoes, and as the brush was so dense that it was impossible to get through with them, we could do little for nearly two weeks, but watch the two slides where we had seen these two bears, and two others that we could overlook with our field-glasses.

About a half mile up the mountain, a little below camp, there was a large slide where the snow, starting from an altitude much greater than usual, had dashed over the edge of a cliff some five hundred feet high, to lie piled up a hundred feet or more against its base. One morning while looking over the slides, we happened to glance at the top of this cliff and, on a little green patch of grass underneath some bushes, saw an old grizzly as white as a goat. For a long time we watched him through the glasses, and that evening he again appeared, and we watched him until dark. He was a fine specimen and looked as though he might be one of those bald-faced fighters that we had come to find, so we determined that if he was there in the morning we would make an effort to get him.

In the morning, sure enough, the bear was there, so we took our guns and started for the cliff. It was a terrific climb. The hill was so steep that in many places we could only proceed by literally pulling ourselves up hand over hand by the bushes. However, after three hours of heart-breaking work, we gained the slide, but only to find our patch of grass unreachable. In fact, the bushes were so high and thick that we could not even see it; nor, guarded as it was by declivities and obstructions, could we be certain where it lay. Fagged and footsore we climbed back to camp. And there, high above us, and serenely beckoning in the sunlight, hung our elusive clearing, and that evening the big white grizzly again came out to mock us. We named him "White Jim," and morning and evening, for nearly three weeks, we saw that bear on this same patch of ground. He had a way of sitting on his haunches near the edge and swinging his head from side to side as though enjoying the view. At last, as day by day the memory of our difficulty grew fainter and the lure of that great white skin grew greater, we could stand it no longer, and decided to make another try, and this time to go prepared to stay all night. So, for the second time, we scaled the all but unscalable mountain, but not a place that was level enough to lie on could we find, and worse yet, just as we got there, a sudden and heavy wind set in. It snowed on the hills and rained down in the bottoms, and, soaked and slipping, we were glad enough to get back to camp with whole bones, and never again tried to steal the pelt of White Jim.

Every day during this time we had seen one or more bear, and had tried all kinds of dodges to get a shot. We cut trails from two directions through the brush to each slide, so that whichever way the wind blew we could creep up against it. We felled foot logs across the streams, and as the high water took them out, we felled more. We exhausted our strength and our ingenuity in ceaseless stalking and planning, and our knowledge of the English language in expressing our opinion of the country and the bears; but the bears were shrewd, and not a bald face among them offered to give us a scrap. And so five weeks after our arrival, satisfied that whatever might be said of their courage, the Selkirk bears were our superiors in cunning, we tied up our tent door and walked twenty-two miles to the settlements, intending to get horses and pack out our belongings. Arrived at the town, we called on an old prospector who was said to have been up this stream, and who had, if rumor spoke true, killed quite a number of grizzlies. Up to this time I had prided myself on being a hunter, and was not a little crestfallen to think that we had been unable to get so much as a shot at a bear. If, I said to myself, this *prospector* had been able to get so many, I should at least have been able to get one small one. We asked him if he had, by clean hunting, ever killed a bear up that creek. He said no, that he had not; he had always baited them and then watched the bait.

We therefore decided to go back and try the bait scheme. We had thought of this before, but it seemed so like taking unfair advantage of the animal that we had put the thought aside. But now our sensibilities were somewhat blunted. We were bent on bear at any price, and we

THE HOME OF THE GRIZZLY—IN THE SELKIRKS

looked around and found a fellow who had an old wind-broken horse which we bought for twenty dollars, and started back for the bear country. It took us two days to get that old horse to the point we wanted it, for every two or three hundred yards we would have to stop and wait until he got his wind; but at last we got him to a spot about three miles above camp, led him across the creek, and killed him at the edge of a slide fifty yards from the creek and fifty feet from a cedar thicket. This, we had found, was a sort of a general converging point for the bear, and we had counted the tracks of twelve that had passed it in a single night. About a hundred yards away we built a blind, so arranged that we could climb up and enter it without going near the bait.

We had already, some distance up the creek beyond the blind, found an old deadfall that some prospector had set up a few years before. Indeed, we afterward learned that the man who said that he had shot the bear had caught them in it. Coleman, who was determined to have one of those bears, whether or no, now proposed that we fix up and set this old trap. We therefore dragged the head and neck of the horse up there, rebuilt the trap, set it, and piled upon the top of it the largest and heaviest log we could roll up. The trap was built between two large cedar trees, so as to compel the bear to enter by the front door, and Coleman rubbed blood over the lintel at the entrance. We also cut off one of the front legs of the horse and dragged it back to camp and placed it on the slide where we had seen the second bear. This slide was about four hundred yards above the creek and we had just got back to camp after placing this last bait, and I was doing some cooking, when Coleman rushed in and said that White Jim from the high cliff was after the bait that we had just placed. We had been careful to place it so that it could be seen from camp, and when I hurried out, there, sure enough, was our old white bear, who had evidently concluded to try another slide for a change, eating grass, digging roots, and nipping off buds, not over fifty yards from the bait. He worked up almost to the bait, but never so much as sniffed at it, and soon most provokingly turned into the brush and disappeared. Later the bears actually fed all around this bait, but not one ever touched it.

We kept a sharp lookout on the old deadfall and on the horse, but for four days nothing touched either bait, although we saw several bears on the slide and their tracks indicated that they had passed close by it. One morning I went up to the old horse's carcass and saw that a bear had made his breakfast from a portion of the shoulder, and so we arranged to keep watch that evening. We did not expect that the bear would return that soon, but we were taking no chances.

During the entire time we had been at this camp we had never seen a bear out later than ten o'clock in the morning, and none had ever come out before two in the afternoon, so we figured that it would be useless to watch between those hours. That evening we watched until dark, but no bears came. The following morning we reached our blind a little after daylight, and finding that nothing had been disturbed, we waited there until ten o'clock and then went to camp. We did some cooking, cut some wood, and returning to the blind at two o'clock found to our amazement and chagrin that the horse had vanished

—lock, stock, and barrel. An old grizzly had come during our absence and had dragged the carcass into the cedars.

As it was quite beyond our combined strength to drag it back, we went to camp, got a rope, and fastening this to the feet of the horse, turned the body over. We then swung the carcass around and, with the rope, once more turned it over. And by repeating this operation we finally got it back to its original position. We now determined to keep an unbroken watch as long as there was light enough to see by, and as it was near the first of June, and there were only about four hours of real darkness, we had to leave our beds soon after midnight, get our breakfast, and climb the three miles of trail by lantern light, in order to reach the blind by the time it was light enough to see our bait with field-glasses. We would then watch until about nine o'clock in the morning, when one of us would go to camp, do the cooking, get a lunch, and return about noon. Then the other would go away and return about three o'clock, when both would watch until it became so dark that we could not see the bait.

For seven days we kept this up, and a good part of the time it rained and was very cold, so that we had to carry up blankets and rubber ponchos to wrap about us. When thoroughly chilled, we took turns stealing down trail to a hill behind which we would run back and forth until warmed up, and all this time the bear never came, except at night when it was so dark that we could not see to shoot, and every time he came he dragged the remains of the old horse to the cedars, and then we, in turn, would drag it back again. And thus, finally, there remained nothing but polished bones.

All this time nothing had happened at the deadfall. After there was nothing more to pick from the horse's bones at the blind, we came to the conclusion that we had had enough, and as we were now out of supplies, we decided to take a last look at the deadfall, spring it, walk out and get the pack-horses, and give it up until the next year. Coleman, as a last attention, had placed the bones in a neat pile, and the next morning on our way up the cañon we were astonished to find that the bear had again been there and that the bones were gone. We went on up to the deadfall, and here another surprise awaited us. The bear had been there also. He had climbed up by the back way, had torn out the logs, had hurled them to the four winds, had taken the head and neck of the horse, and had left for parts unknown. When Coleman looked at the wreck, he sat down on one of the logs and, rubbing his head, said, "Let's go home. We are babes in the wood, compared to these bears." So once more we fastened up our tent door and took our guns and walked out of the country. We had seen twenty-one grizzlies and had never fired a shot. We had only once been near enough to shoot at one, and that was the first one we had seen. So we went home to think it over, and plan and scheme until another spring, when we were determined to try it again.

XIII

THE SELKIRKS REVISITED

THE next spring—for we held to our determination to again match wits with the Selkirk grizzlies—we reached our old hunting grounds during the first week of May, Mr. G. O. Shields being also of the party. This time, however, in pursuance of a plan settled upon during the winter, we took along three dogs, purposing to see what could be accomplished by running the bear. We hired these from a man who bred them for running coyotes and cougars, but who guaranteed them to run "anything that wore hair," and we paid fifty dollars for the use of the three for the trip, agreeing that if we wished to keep any of them, or if any of them were lost, we would pay one hundred dollars each for them. They were a cross between the foxhound and the bloodhound, and had good speed, being able to run down a coyote in a few hours.

When we arrived at our former camp, we found that an old trapper had already moved in on snow-shoes and had about a dozen traps set. He had not yet caught any bear, although he had seen two the day before our arrival, and, as we did not care to run hundred-dollar dogs among set bear traps, Coleman decided to buy the old fellow out and have the field to ourselves. It took two days' talking to strike a bargain, and then it was agreed that, for one hundred and fifty dollars, the trapper would spring his traps and leave the country until we were through hunting, when he was to be at liberty to come back and take up his work where he had dropped it.

Just as the bargain was closed I looked up the slide where we had seen our second bear the year before, and there, six or eight hundred yards away, was an old grizzly unconcernedly eating grass. Coleman and I seized our rifles and started across the valley toward one of the trails we had cut for such occasions, and the old trapper, in a holiday mood, reached for his gun and brought up the rear. We had the creek to cross and a steep bank to climb beyond it, and between the camp and the creek there was at least four feet of snow, its surface made treacherous and insecure by hidden cavities that had melted around the warm trunks of fallen trees. Just as we reached the creek we heard a cry, and turning, saw the old man's face projecting ludicrously from a snow-drift, and looking like a red and angry sun poised at the horizon. It was no time, however, for either jokes or rescues, and Coleman and I crossed the creek, climbed the bank, struggled through a thicket of brush surmounting it, and gained the trail that we had cut the year before; and then, with bent backs and silent steps, we worked our way to the edge of the little stream fed by the melting snow and just beyond which we had seen the bear. Cautiously we covered the last few steps, cautiously we peered between the branches, and (it was so like old times that it made us feel at home) the bear was gone.

As we stood for a time peering carefully about us, my

TRAVELLING IN THE SELKIRKS

eye was caught by something in the bushes just across the slide—something that looked like hair moving in the wind—and I told Coleman that I believed the bear was lying down over there in the sun. He, however, pooh-poohed the idea, saying that for his part he did not believe these northern grizzlies ever lay down anywhere; and he proceeded to jump the creek and start across the slide. As he did so I saw the spot I had taken to be a sleeping bear move, raise its head, look over its shoulder at Coleman, and start to move away; and as there was no chance to give Coleman the shot, I hurriedly fired myself and scored a clean miss. The bear was now well on his feet and under pretty good headway, but I would have reached his shoulder with my second shot had he not, just as I pressed the trigger, turned his head to look at us and taken the bullet full in the mouth, which it swept clean of front teeth on both jaws. Coleman now got into the game, and as the bear was going straight away from him, planted a shot fairly in the centre of his back, cutting the spine, and he rolled down the bank to our very feet. And so, at last, tamedly enough we made our first kill among the reputedly savage grizzlies of the Selkirks.

The bear was not a large one, weighing only about three hundred pounds, and Shields, who, with the old trapper, now came up, said it was not the one we had started after. That one, it seems, had moved off into the bushes almost as soon as we left camp, and examination showed that the one we killed had had a very comfortable bed in the brush and had probably been lying there for some time.

The next day the old trapper made the round of his

traps, sprung them, and taking one or two of them with him, went over the range to our left, to where he said there was also good bear country. We were now anxious to find a fresh trail and see what the dogs would do with the grizzlies, or—for we owned ourselves subject to doubts—what a grizzly would do to the dogs. The following morning, some distance up-stream, we came across tracks that had evidently been made the evening before, and as these led toward camp, and we thought the animal might be near by, we followed them up and jumped the bear within two hundred yards. He immediately plunged downhill into a thicket near the creek, and leaving Coleman to watch the place, I went back to camp to fetch the hounds. On our return I found that the bear had not come out, and when the dogs saw his track in the snow they stopped, took one good sniff, and then gave tongue in a way that made the cañon sound like a hurdy-gurdy.

But the bear at the very start had a surprise to spring on us. The water in the creek was bank high, and so boiling swift that we thought nothing could live in it for a moment. We had therefore taken it for granted that the bear, when the dogs dislodged him, would have to run either up or down its banks or toward the hill. Instead of this, however, he made straight for the creek, plunged in, and had crossed before we could come within shot of him.

A moment later the three dogs, baying their best, plunged into the current and were instantly swept downstream at race-horse speed; while we stood open-mouthed and mentally entered up three hundred dollars for drowned dogs on the debit side of the ledger. But the dogs were game. Swimming desperately, they at last, far down the creek, managed, one by one, to make the shore. And then, giving but a moment to shaking themselves free of cold water and bad memories, they once more gave tongue and took up the pursuit. Coleman and I, in order to cross, had to go down-stream quite a distance to our foot-log, and by the time we had retraced our steps the chase had swept out of hearing. We followed the trail for half a mile or so and found that the bear was making straight up the side of the mountain, and thinking it likely that he was heading across toward the little bottom where we had killed the horse the year previous, and where so many trails converged, we decided to skirt the base of the hill instead of scaling it. But when we reached the bottom we found no trace of the bear, nor could we hear the dogs, and as Coleman by now was nearly tuckered out bucking the deep snow, he decided to return to camp.

But for my own part I had no such intention. Seeing the first part of the bear's trail had fired me to see more of it, and so, parting from Coleman at the point where the grizzly had left the water, I followed in his wake. For some distance he had kept alongside the hill, and then, finding that the dogs were nearing him, he had turned straight up the mountain. The surface of the snow was soft and his tracks told their story plainly, and he was a wily old campaigner, and a strategist that knew his own advantages. In going up the mountain he had deliberately picked the worst possible going, and, in places, had even scaled cliffs that the dogs had had to go around. Again, he had turned at right angles and run horizontally along hillsides so steep that the dogs, to keep from falling headlong, had had to give ground and veer downhill. He accomplished this feat and kept his huge weight from slipping downward by simply twisting his fore paws till his huge claws engaged the snow like hooks. To put it plainly, when he was headed south, his fore claws pointed east, uphill.

What with these tactics, and his natural speediness, he was so far ahead of the dogs when he reached the summit of the mountain that, as his trail plainly showed, he had stopped and stood on his hind legs to look or listen for them, and then, dropping again on all-fours, he had started down the opposite slope. Now, if his tactics in the ascent had been masterly, his method of going down was spectacular. The snow was comparatively soft for a depth of two feet or so, and that bear's trail looked like the track of a huge boulder. He had simply turned on full steam, pulled the throttle wide open, and let her go. There were places where I took nine steps to cover one of his jumps.

Arrived at the bottom of the mountain I found that, after all, the bear had made for and passed the place at which we had looked for him, and that had we waited we would probably have got a shot at him. As it was he had followed one of our cut trails for half a mile and then again turned up the mountain; and as it had begun to rain and I had no coat, and as no man could successfully follow a bear afoot in that snow, I also decided to return to camp.

But fortune is a capricious lady, especially in the wilderness. She takes the keenest delight in making game of our best efforts and in mocking our pet vanities, and then she tosses the unexpected into our laps. Just as I turned back a large grizzly came breaking through the brush at my right, straight into the slide where I was standing, and, without moving a step, I dropped him with a single shot.

He was a fine animal, much larger than the first one we had killed, and after looking him over I left him where he had fallen, and we came up the next day and skinned him.

The dogs did not return to camp for about nine hours, and while we never knew whether they caught up with the bear or not, we had our own opinions. I doubt, indeed, if, master of strategy as he was, and with the lead he had when I turned back, the dog ever lived that could have caught him, or, for the matter of that, any grizzly in those hills that felt disposed to run. As we were to discover later, however, by no means all of them were so disposed. But the alternative that they adopted and its effect upon these particular dogs was a puzzle we were a long time in solving.

Meanwhile our enthusiasm for hunting bears with dogs suffered an eclipse. It seemed to us that the bear and the dogs had all the fun, and we could not see where our chances of a shot came in. So we tied the hounds up in camp and reverted for a while to our original tactics.

Toward evening the next day we climbed up to the big slides where we had once set the old deadfall, and sighted two grizzlies feeding along the edges of the bushes. When we had succeeded in creeping up to within three hundred yards of them I thought it best to risk a shot; but Coleman decided that, as they had not yet seen us, we could get a little nearer, and although one of them stood up and sniffled over his shoulder, when he dropped down, and both went on feeding their way toward a clump of brush, Coleman started to cross a small creek in order to approach them. As he did so we saw the bears, now on the opposite side of the brush and two hundred yards further away, sprinting for all they were worth for the mountain. We at once began to shoot, but though we saw dirt fly up, now in front and now behind them, we failed to hit them, and they soon veered to the left and disappeared up a gulch, while we returned to camp, promising each other to be more prudent another time.

We had already seen the tracks of the old grizzly that had eaten our horse the previous year, and that we had nicknamed "Big Foot." He had evidently made several trips to the old bones, and we decided to make a try for him in the morning. After breakfast, therefore, we walked up to the slide and, as luck would have it, the first thing that we saw was our old friend about four hundred yards away, digging roots and browsing in the open ground. We immediately sat down and held a *pow-wow*, the subject of debate being whether we should shoot from where we were or try to get nearer. The experience of the previous day finally decided us and, taking careful aim, we both fired at the same time. Big Foot jumped about five feet into the air, turned his nose up the hill, and beat his own best record into a clump of juniper bushes. We thought we had wounded him and I was about to go to camp and get the dogs, when we saw him appear far up the side of the mountain on a boulder that overlooked the valley. And when I planted a ball against the rock directly under him, he turned a complete back somersault and, landing in a cloud of dust among the bushes, disappeared. This time we returned to camp as disgusted with prudence as the day previous we had been out of humor with rashness.

We now thought we might get a bear by baiting, so we took the body of the grizzly we had shot, roasted it, and

placed it where the old horse had been killed the year before. But the bears did not touch it.

The old trapper came over to see us and reported seeing
quite a number of bears in his part of the country. He had
wounded one, but it got away, and he had also set several
traps, but none of the animals could be coaxed to walk into
them. He had used fish, honey, and molasses and had even
killed wild goats for bait, but the bear would have none of
them. After he had gone back, however, having killed a
goat ourselves, we laid some pieces of meat near the bones
of the old horse. We had also tied a choice piece for our
own use to the end of a rope, and had swung it from a high
limb for safe keeping. The next day we found that the bait
had not been touched, but the piece we had swung from
the tree was gone. The tree was clawed and torn, the rope
was broken, and our goat ham had vanished.

This was almost too much for Coleman. He said he
had tried it the year before, and he was satisfied that no
one was smart enough to get ahead of these northern grizzlies, but I could not for the life of me understand how a
grizzly got up that tree. I would not, indeed, believe that
a bear had stolen the meat, and so proposed to set a trap,
not, however, with the expectation of catching one. We
therefore took one belonging to the old trapper and set it
at the foot of the tree, tied the rest of the goat meat to the
rope, and again swung it in the branches, and the next
morning we had a very large brown bear. This was the
first bear other than a grizzly that we had ever seen in that
country, and we were, of course, sorry to have caught it.
It was the first bear I had ever caught in a trap.

We were now at our wits' end and ready to follow up

any hint that the dogs gave us. We had turned them out one morning for a few minutes' exercise, and although startled when they immediately bolted howling across the creek, when they stopped barking after running a few hundred yards, we paid no further attention to them; but when they did not come back by the time we were through lunch, we decided to go after them. The truth is that we were rather uneasy because, when the old trapper had left, he had told us of one trap that he had forgotten to spring, and we had promised to attend to it; but it had slipped our minds, and now that the dogs were out we became worried about it. He had told us where it lay, and I started to go up and spring it, but as I crossed the creek I stopped about half-way over to listen, thinking I heard the dogs, and just as I halted, the bark on the log gave way and I shot down, striking my side heavily on the log and breaking two ribs. After climbing out of the creek and getting my breath, I dragged myself up to the trap and sprung it, and then got back to camp, and when Coleman came in we took some flour sacks, made a corset for my broken ribs, and then sat down and waited for the dogs. They did not return, however, until some time in the night, when they came in and stuck their cold noses in our faces. We tied them up and returned to bed, deciding to go up-stream in the morning and find out what they had been doing.

After an early breakfast we started, and not over two hundred yards from camp we saw bear tracks that measured eight by twelve and a half inches and dog tracks par-alelling them on either side. This bear, evidently, had not run; he had crossed the creek and walked steadily along up the trail. Every little while we could see where he had turned and struck at the dogs, and where they had jumped aside, but only to come back when he moved on. But why we had not heard them bark was a mystery to us, and this only deepened as we followed the trail for about three miles and found that during this entire distance the bear had not gone out of a walk. Of course we felt that we had lost a fine bear.

We decided, therefore, to try the dogs again.

There was a slide across the creek from us that we had named the Gateway. It ran through a deep gulch in the face of the mountain, and opened out suddenly into a fan-shaped declivity that sloped to the creek nearly half a mile below, and was covered clear to the bottom with fifty feet of snow. Above this opening, the mountain on one side showed steep, high cliffs, broken into terraces. On the other grew an open forest with little underbrush, but much down timber, and just beyond, round a point of the hills, was a mountain stream that fairly hurled itself down to the creek. The whole mountain was so steep that we could only climb it by grasping at bushes and fallen trees.

Just as we came opposite to this slide on our next excursion, we missed the dogs and, turning to look for them, heard a chorus of barks receding up-stream on the other side of the creek. We followed, but after going half a mile, the dogs turned and made straight for the Gateway, and when they suddenly stopped barking, determined to discover the meaning of such conduct, we crossed the stream and found them sniffing about the edge of the mountain torrent, the bear evidently having taken to the water and put them off the scent. We waited some time, but the dogs seemed unable to pick up the trail, and as my broken ribs

incapacitated me from much climbing, we gave it up. The
next day we tried the Gateway by ourselves. When about
half way up the slide, we saw a big grizzly some hundreds
of yards away, moving along as though time were no object to him. Coleman tried to get within range, but when
he had gone fifty yards or so the bear turned and, still with
every appearance of leisure, disappeared behind some
bushes. After some waiting, as he did not reappear, we
went to investigate, and found that, as soon as he had got
behind the bushes, he had started on a dead run. I then
went back to camp for the dogs, and when we put them
on the trail, they promptly followed it round the point to
the torrent and again lost it.

The first of June, the date when we had ordered our
pack-horses to be sent in for us, was now approaching.
The last week of May, Dr. C. S. Penfield, of Spokane,
joined us, and we continued to hunt every day, but with
no better success. At last we had but one day left before
we were to go out, and we decided to take the dogs and
look up our old friend Big Foot. We had no trouble in
jumping him, but, as usual, he easily threw the dogs off the
trail, and so, as a last resort, we thought we would give the
old fellow in the Gateway a final try.

We laid very elaborate plans. The doctor and Coleman
were to stand at the main creek, one above the slide and
one below it. I was to take the dogs across, let them loose,
and then hasten to the torrent around the point of the
mountain, and lie in wait to see what it was the bear did
to throw the dogs off his track. Almost as soon as I started
up the slide, the dogs struck a scent and headed up-stream.
This was something we had not counted on, as they had

nearly always, after circling about in the thicket, made for the point. It was useless, however, for me to try to change the programme, so, for lack of anything better to do, I kept on toward the Gateway. The dogs swept on up-stream and soon passed out of hearing, and I kept on up the slide.

As I approached the point where the dogs had always lost the trail, I saw that an old she bear and two cubs had been digging there for roots that morning, it having rained nearly all night and the tracks having been made since it stopped. I then climbed the high snow banks and listened for the dogs, and soon I could hear them, apparently returning. On they came nearer and nearer, while I, with a clear view for two hundred yards in every direction, stood with my gun half raised, waiting for the bear to break cover and make for his favorite place at the point. I was certain that this time he was making his last run. But when the dogs appeared, going like mad, there was not a solitary thing in front of them. They passed just below me and, when they reached the bank of the stream, turned suddenly to the right, ran up into the point, and as suddenly ceased barking.

I watched and waited, and they soon came out and ranged about—interested, but not at all excited—where the bear and cubs had been digging. I was now utterly at sea and thoroughly determined to find out what had become of that old she bear and her cubs, and why the dogs should run so fast and true to this point and then lose the trail. It seemed certain now that this bear was the one they were after, and that they had simply followed her in her various meanderings up the stream and back again. I therefore whistled to them and started to crawl through the thick brush into the open timber beyond. My broken ribs were not yet healed and I had my gun in my hand and a camera on my back, and as it was hard work, I made no effort to get through without noise. The last thing I thought of was that that bear was in there.

Upon reaching the farthest edge of the tangle I came to a fallen tree that lay about four feet from the ground, with the brush so thick underneath it that it was necessary to climb over it. Catching hold of a limb with each hand, I drew myself up, painfully enough, till my knees rested on the trunk, and there, not over fifty feet up the hill from me, with her fore paws planted on another log, I saw a grizzly that, just then, looked to me to be about the size of a cow. I made a frantic effort to get on my feet, and as soon as she saw that it was not the dogs that were after her, the bear started pell-mell up the hill. I had no time to stand up and aim, so I fired off-hand from my knees and then quickly grasped a limb to keep from pitching forward on my nose. I hit the bear somewhere, for she rolled back against the log, but, almost instantly recovering herself, again started up the mountain. I took another off-hand shot and again grabbed the limb to keep from falling. Again the bear rolled back, started off again, and for the third time I sent a shot after her; but this time she kept on up the hill, and the last I saw of her she was disappearing round the bluffs into the timber with those fool dogs now yelping in pursuit.

And then the solution of our mystery suddenly dawned on me. The bear had not fooled the dogs at all. The dogs had fooled us. They were guaranteed to "run anything that wore hair," and as long as the thing that wore hair would run, they were as good as their guarantee. But there their responsibility ended. When they struck a bear trail they followed it, yelping with anticipation; when they jumped the bear, if he ran, they ran after, still yelping. If he refused to run and contented himself with walking, they followed in silent hope of his changing his mind. But if he entered a thicket and stopped, they took it as a sign that there was nothing doing, and came home. This was why we had heard no sign from them the day they had escorted the bear three miles up the trail from camp. This was why they had always stopped at the Gateway, where the bear took refuge in the down timber. This was why Big Foot was still living up Wilson's Creek. I sat down and laughed. Then I took up the bear's trail and found such a showing of blood as to make me believe that she lay dead not far away. But it gave me so much pain to pull myself up the hill that I soon had to give over, and as it set in to rain that evening and kept it up all night, the blood was washed away so that we never found the bear. We did take the dogs and go after the cubs, and we jumped them at the point where I had shot the mother. I saw one of them as he ran with the dogs after him, but these stopped barking as soon as he stopped running, and after a couple of hours we gave it up. It was impossible to get the cubs out of that brush with those dogs. That night the horses came, and the next day we packed up and returned home.

Up to this time I had never doubted that a she grizzly would fight for her young. I would have staked almost anything that an old grizzly with cubs would charge everything within fifty yards. Here, however, was one that I got within fifty feet of and wounded, and yet she scuttled off and left her cubs to look after themselves. One other instance I saw, a year later, at the same place. I was hunting with another party when we saw from the opposite side of the creek an old grizzly and three cubs come out on the snow, walk across, and go to digging roots. We began a stalk, and when some two hundred yards away, they got wind of us and started back. My companion and I took several running shots at them, and one of the cubs was slightly wounded across the back. The four fled up the trail where I had chased the wounded bear the year before, and in this trail our old friend the trapper, who was in there again, had set a trap, and the wounded cub ran right into it. He at once, and very naturally, began to bawl, and we fully expected the old bear to charge us. As she did not, I went up and forced the cub to continue his bawling while the other man stood, with his gun ready to shoot when the mother appeared. But she and the other cubs continued their retreat and left the bawling cub to his fate.

XIV

THE UNEXPECTED

IN hunting, as in other matters, it is more often than not the unexpected that happens. The actual earned runs in the game, especially in the game of grizzly hunting, are few and far between. To make one of them requires a varied knowledge, the skill that comes from long experience, dogged perseverance, and an infinitude of patience. And the memory of success is a joy forever. But the candid hunter, when he looks backward, will own that, for the most part, the grizzlies he has worked hardest for are the ones he never got, and that more opportunities are *grasped* than *made*.

Some years ago Mr. M. W. Pope, of Baltimore, and myself were hunting wild goats in northern British Columbia. This was virtually the only game to be found in that part of the country, and there were none too many goats. There were, it is true, a few sheep, but they were *so* few that we did not consider them; and as for bears, there were practically none.

I had but recently spent three months in the same region, trying to photograph game, but with little success. At a large lick I had seen about twenty goats, and in another valley a few sheep, but I did not see a single deer or bear, and only two or three bear tracks.

Mr. Pope joined me the last of August, and we worked our way back to where I had seen the goats, this being the game we had come to hunt. The animals, however, had moved, and we decided to go farther up toward the divide between the Saskatchewan and the Athabaska Rivers, and finally camped on the Saskatchewan side of the divide, not far from the summit.

The next morning we saddled two horses and rode up toward the crest, some three miles away, and when near the point where we would have to leave our horses, I looked up to our right and, on top of the ridge above the timber line, saw quite a large grizzly, running for all he was worth. I called Pope's attention to him, and as he was entirely out of range, being some eight hundred yards away, and as we supposed from his rapid flight that he had seen us, we sat quietly on the horses and looked at him. The horses, too, saw the bear and they also watched him. For two hundred yards or so he continued his flight, and then, to our amazement, he turned down the ridge and came straight toward us on the jump. This was another story, and, while we both dismounted, I held the horses by their heads so they could not make any disturbance, and Pope stepped a few feet ahead and dropped on one knee ready for a shot in case the bear came near enough. He was armed with a .45–70 rifle, while I had only a Stevens .38 shot-gun for shooting grouse.

The bear came on downhill at the same mad gait until he had covered half the distance and was not over four hundred yards above us, when he suddenly dashed into a little thicket of fir bushes and disappeared. As he did not come out again we went into a committee of the whole

HE LANDED IN THE LITTLE TRAIL THAT WE WERE FOLLOWING

to discuss our chances of crawling up on him. We were sure now that he had seen us, but there was not a bush between us and the bear, and there was nothing we could tie the horses to, and we did not dare leave them for fear they would run and frighten the bear. At this stage of the discussion, however, we saw the bushes sway, out jumped the grizzly, and down he came again straight toward us. It seemed as if he must surely have seen us, but I told Pope to let him come as long as he would, and he actually came up within a hundred and fifty yards of us, when he turned to the left, stopped, and commenced digging out a ground squirrel.

This was our chance. The bear was standing broadside on, and Pope fired, hitting him just back of the shoulder and piercing the heart. It was a good shot. The bear turned a somersault, cart-wheeled down toward us, and never stopped until he landed in the little trail that we were following, and not more than thirty yards from where we stood.

This was Pope's first grizzly, and from the stories he had been told of how wild these bears were and how hard to kill, he had felt that it was doubtful if he ever got one. Yet we, with the two horses, had been standing all the time in plain view. Not a bush screened us, and the horses kept their ears pointed forward and watched the bear from the time he left the ridge until Pope shot him. It was a strange adventure.

A year or so later, while photographing in the Bitter Roots with Mr. W. E. Carlin, we spent most of the summer and early fall on one of the divides between the South and Middle Forks of the Clearwater River. We were photographing small birds and animals, and were only doing enough hunting to supply the larder.

Once, when we had been without meat for a couple of days, Carlin, who was not feeling well, urged me to go out alone and bring in some game. I wished him to go with me, however, and, as an inducement, suggested that we take the horses and go some twelve miles to the west over to a large marsh and hunt for small moose. After some coaxing he consented, and we got the horses and rode over to the marsh. This was always a fine place for moose and grizzlies, and I have seldom been there without seeing one or the other, and more often both, but we were not thinking of anything but moose at the time.

The marsh was two or three miles long and from a few hundred yards to a mile in width, with several little ponds scattered about it and a small stream running zig-zag through it. It was mostly covered with small birch brush from two to six feet high. And there were little open parks here and there, in which grew a plant much relished by the grizzly. Arrived at the marsh we saw a good many fresh tracks of moose, and also noticed where an old grizzly with two cubs had been working among the plants; and as the bear signs were of different ages, we concluded that this family had been spending the summer there. We worked slowly up the stream, but saw neither moose nor bear, and the farther up we went the less sign we saw of bear and the more of moose.

After a time Carlin said he felt ill and would have to sit down and let me go on alone, but I persuaded him to go with me to a little bend in the creek, promising that if we saw nothing then he could wait and I would try and get

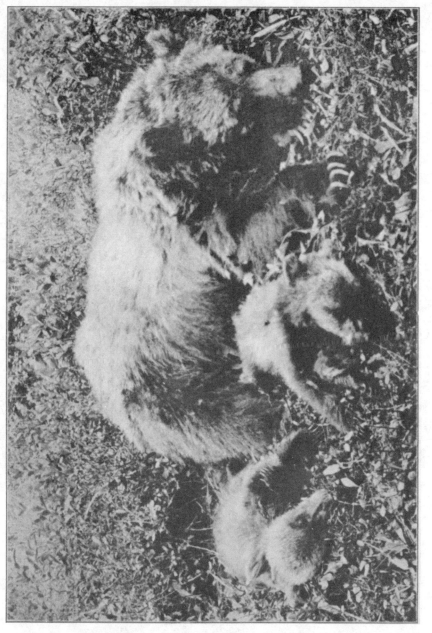

IN THE BITTER ROOTS—THE OLD GRIZZLY AND HER TWO CUBS

the meat. He went on then, in a half-hearted way, and we had gone perhaps three hundred yards, when, on looking across the creek and behind some bushes among a few fir-trees, I saw what I took to be a moose. I had scarcely called out, "Look at the small moose," when the moose suddenly stood up on its hind legs and looked over the bushes at us.

Carlin said, "It's a bear," and gave her a .30–40 bullet through the shoulders, and she reversed ends so quickly that we could not afterward remember seeing a motion until we saw the bottoms of her hind feet two yards in the air. Then she struck the ground with a loud roar, and two cubs joined in with their bawls. We sprang across the creek, found the old bear down and moaning fearfully, and Carlin planted another bullet in her head that shattered it into many pieces.

The cubs were in the bushes and they now set up a roar. One of them showed itself and I laid it out, and about the same time Carlin got a shot at the other and ended the encounter. This was the first time that either of us had had the opportunity of observing the effect of high-power bullets on living targets. When we skinned the old bear, notwithstanding that no shot had hit her back of the shoulder, we found that the bottoms of her hind feet were clotted with blood under the skin, presumably owing to the shock.

The excitement cured Carlin of his illness, and in spite of the fact that there was no meat for dinner, we returned to camp in high spirits.

XV

A SPRING-GUN AVOIDED

THE grizzly bear far excels in cunning any other animal found throughout the Rocky Mountains, and indeed, for that matter, he far excels them all combined. I have seen many and various examples of his shrewdness, but never, perhaps, a more striking instance than one which took place while I was hunting with a party in the Bitter Roots. We had been trying to catch a grizzly alive, and had been using for the purpose a bear trap with ropes stuffed into the corners of the jaws, so that these would only come together far enough to hold the bear's foot without inflicting an injury.

We were camped on a large branch of the Clearwater River, and one of the party having killed an elk, we dressed it and carried the trimmings to a point about a mile from camp. Here we had built a pen of logs and in this we planned to set the trap, attached to a light log so that the bear would not pull his foot out in his struggles to get free. On the second day I went to examine the trap and found the pen demolished, the bait taken out, and everything that was movable piled on top of the trap. Fifty feet away I saw a large pile of moss and leaves scraped together and beside it a bed, where the bear had been lying. I kicked open the mass of leaves and found the remains of the bait hidden there. I then dug the trap out from the wreck of the pen, set it, and slid it in among the leaves and moss, throwing some meat around it, and that night the bear came again, picked up all the loose meat, but touched nothing under the moss.

A mile up the stream there was an open hill, in the side of which there was a large elk lick where three of our party had been watching, hoping to bag an entire family of elk that frequented it. The morning after the bear made his second visit to the trap this family of elk, a bull, a cow, and a calf, were killed on a small flat above a stream that entered our river near the lick. This stream ran through a deep gully and at the point where the elk were killed the sides of this gully formed a sort of shoot, sloping at an extremely steep angle, from a point a hundred feet above the water, sheer to its edge. At the north end of this shoot a mass of down timber, their ends sticking in the water, formed an effective barrier. At the south it was bounded by an outcropping of rock which fell, in a succession of ledges three or four feet high, to the water in front and to the shoot at the side. Twenty-five or thirty feet from the creek, in the middle of this shoot, grew an old cedar-tree.

Of the three animals that had been shot, we dressed the calf and took it to camp, all except the entrails; we saved the head and hide of the cow, together with the loins and the other choice cuts, but of the bull we took only the head and the hide. All that was left we dragged to the top of this shoot and rolled down it. The carcass of the cow happened to hit the tree and lodged against its trunk. The old bull brought up at the bottom. And there we left them.

The next morning one of the party happened to pass the place and found that our grizzly had been there, had made a meal off of what was left of the cow, had then gone down, seized the carcass of the bull, dragged it up the steep hill, and placed it on top of the cow. He had then gone down again, gathered up the remains of the calf, added them to the pile, and, digging into the side of the hill, had buried the lot.

On the opposite side of the creek the bank was equally steep. At its top there was a heavy growth of timber and underbrush, and we found that the bear had come from this cover, and that his trail, winding down the bank, passed between two large cedar-trees on his way to the creek. We therefore set the trap between these two cedars and watched the place until dark, but the bear did not show himself. The next morning, however, we found that he had been to his cache for a meal, but that, instead of travelling his old trail between the cedars, he had circled them, and although we watched again that night, we saw nothing of him.

We were now within two days of the expiration of our stay, and some members of the party, who had never seen so large a bear (his track measured fifteen inches), determined to set a spring-gun for him. After some planning they settled upon the following scheme: A few feet down-hill from the old cedar tree, behind which the bear had buried the meat, they rigged up a gun with its butt against the base of the rocks, and its muzzle pointing toward the barrier of down timber. A silk fish-line was then attached to the trigger and fastened at the other end to one of these fallen trees. Thus, if the bear approached his cache from

From a photograph, copyright, 1909, by J. B. Kerfoot

HIS SUSPICIONS AROUSED

below, he would of necessity run into this stretched cord, and receive the bullet in his side. He had, heretofore, always approached from this direction, but in order to make assurance doubly sure, another gun was rigged above the tree, with a line stretched parallel to the first and about twelve feet from it.

Having completed these arrangements we returned to camp, but though we slept with one ear open, the expected report was never heard. The next morning when we went out to examine our trap we found written in footprints on the dirt as wonderful a record of animal sagacity as I have ever seen.

The bear had come as usual for his evening meal. He had come down from his covert, circled the two cedars where our trap still waited for him, crossed the creek, and climbed to where the lower string was stretched across his path. But though he had come up to it he had not touched it. On the contrary his tracks showed that he had turned to his left, followed the string to the barrier of fallen trees, had found himself unable to get around it there, had turned and followed it to the rocks, had found himself blocked there also, and had retraced his steps to the creek. He had then circled the rocky point, had climbed to the flat above, and had tried to reach his cache from the other side. But here he had again encountered the suspicious string. Once more he followed it to the down timber, turned and made his way along to the rocks, and then the wily old fellow had climbed out on to the rocky point and, making his way from ledge to ledge, had arrived safely between the two strings, eaten his meal in comfort, and gone out the way he came. We never got that bear.

XVI

A PHOTOGRAPHIC EXPEDITION

A GREAT many years ago my interest in natural history, which grew out of my interest in hunting, caused me to give a certain amount of attention to photography. Little by little, as I became more expert in this, I took to carrying a camera with me on my various expeditions, and finally I came to making excursions with no other end in view than the photographing of game. It was a long time, however, before I developed a definite ambition to photograph a grizzly, because the difficulties which presented themselves in that field were so many that at first I saw no way of overcoming them.

Much hunting has not only made the grizzly very shy, but has caused him gradually to become even more nocturnal or, to be accurate, crepuscular, than he was originally. It follows that in these latter days the chances of obtaining a daylight picture of a grizzly are almost negligible, and though by some lucky chance one might meet a bear in a snap-shotting light when one had a camera ready, the coincidence would be too unlikely to depend upon. When, therefore, I began to think seriously of attempting to photograph these bears, I of necessity turned my mind to flash-light, and for several years I worked and experi-

mented to that end.   The most favorable time to operate
being between sundown and dark, it was impracticable
to set up a camera and leave the lens open and provide for
the exploding of a flash when the bear came along, and I
therefore set about perfecting an electrical device which
at the same time would explode the flash and spring the
shutter of the camera.   My first idea was to have this
apparatus operated by the bear himself, and to that end
I constructed it so that the trigger could be tripped by
pressure applied to a fine thread or wire, which could be
stretched across the trail; but though I soon succeeded in
getting this mechanism to work well at home, actual
practice in the field developed a succession of difficulties
which had to be overcome little by little, and as field trials
were scarce and expensive, it took me a long time to arrive at satisfactory results.

By the time my camera was in working order, the bears
on which I had expected to use it were all but things of the
past; and having heard for a number of years that the
grizzlies of the Yellowstone National Park had become
comparatively tame, and that it was no difficult task to
photograph them, and having hunted grizzlies in all the
country round the park without finding the bears there
different from what they were in other parts of the country,
I determined to take my camera to the park and study the
grizzly in this field.   This was in 1906.

I was armed with a permit from Major Pitcher, the
acting superintendent, which allowed me to photograph
and study the grizzlies, provided I did not molest them in
any way.   I went first to the Grand Cañon.   I found
there quite a number of grizzlies feeding in the evening at

the garbage dump back of the hotel, and for a few evenings I watched them there in order to determine the direction from which they came, and to ascertain how many were using this feeding ground. After watching for a few evenings I found that there were about thirty grizzlies all told that came there. There were several old she bears with litters of cubs, several litters two and three years old that had left their mothers, but were still running together, and several old fellows that came and went by themselves.

While I was watching the dump in the evening, I travelled the surrounding country by day to see if any of these bears could be seen by daylight, and though I scoured every thicket and gully, not a grizzly did I thus see during some two weeks' sojourn there. In this respect they were much more timid than they were in a great many places throughout the Selkirk and Rocky Mountain ranges.

My next move was to find out where these bears hid when they were not feeding, for I have never yet seen a grizzly that did not have a home, either in some dense thicket or in some heavy timber or in some high mountain. I followed some of the more travelled trails for several miles and found that nearly all of these grizzlies had their headquarters in the range of mountains around Mt. Washburn. I then selected their largest highway, and after setting up my camera, concealed myself one evening about a hundred feet from the trail and to leeward of it, and watched for the coming of the grizzlies. Across the trail I had stretched a number forty sewing thread, one end attached to the electric switch and the other to a small stake driven into the ground beyond the trail. Just below where I had located, there was an open park in which the bears had been feeding, as was shown by the grass that had been nipped and the holes that had been dug for roots.

For some hours I waited in the bushes and fought gnats and mosquitoes. I saw several black bears pass along the hillside, but not a grizzly showed his nose until after the sun had set and the little marsh in the park was covered with a mantle of fog. Suddenly then, far up the trail, appeared what at first looked like a shadow, so slowly and silently did it move. But I knew at once, by the motion of the head and the long stride, that a grizzly was coming to the bottom for a few roots and a feed of grass.

I watched closely to see if he acted differently from bears elsewhere that are supposed to know less of man. I could not, however, detect the slightest difference in his actions from those of bears that had never seen Yellowstone Park. All his movements were furtive and cautious, as if he expected to meet an enemy at every step. He would advance a few feet, and then stop, turn his head from side to side, scent the air, and peer in every direction.

I was, of course, very anxious to see what he would do when he came to the thread across the trail, and I had not long to wait, for he came on steadily but slowly and, when within ten feet of the thread, he stopped, poked out his nose and sniffed two or three times, raised up on his hind feet, took a few more sniffs, and then bolted up the trail in the direction from which he had come. This bear did not seem to have been very successfully tamed.

A few minutes after he had gone three more appeared. These were evidently of one litter and appeared to be between two and three years old. They came on with the same cautious movements, and when they were close upon the thread, they also stopped and went through a similar performance. The one in front pushed out his nose and sniffed gingerly at the suspicious object. Those in the rear also stopped, but being curious to learn what was causing the blockade, the second one placed his forefeet on the rump of the one in front, in order to see ahead, while the third one straightened up on his hind legs and looked over the other two. They made a beautiful group, and just as they had poised themselves, the one in front must have touched the string a little harder than he had intended to, for there was a sudden flash that lit up the surroundings, and I expected to see the bears go tearing off through the timber, but, to my utter surprise, nothing of the kind happened. They all three stood up on their hind legs, and looked at each other as much as to say, "Now, what do you think of that?" and then they took up their investigation where it had been interrupted, followed the thread to where it was fastened to the stick, clawed up the spool, which I had buried in the ground, sniffed at it, and then went back to the trail, where they had first found the thread. Here they again stood up, and then, having either satisfied their curiosity or becoming suspicious, they turned around and trailed away through the timber. As far as I could see them they went cautiously, and stopped at frequent intervals to stand up and look behind them to see if there were any more flashes or if anything was following them. Unfortunately this picture was utterly worthless. I had failed to use enough flash powder, and when I came to develop the plate, it showed only the dimmest outline of the animals.

Soon after this an old she bear with three cubs came

down the trail, but they were just as cautious as the others had been. Every few feet the mother would stop and sniff the air, and the cubs, fascinating little imitators that they are, had to copy her every move. If she stood up on her hind feet, they also stood up on theirs. If she stopped to sniff the air, they would run up and, placing their tiny feet against her sides, would peer wisely and anxiously ahead, until the old lady started on again. When she came to the thread she stopped short, and while she was making her investigations the cubs stood with their forefeet against her and awaited the verdict. It was sudden and apparently surprising, for, after satisfying herself that the obstacle was placed there for no good, she gave a lively snort that could have been heard for a hundred yards, and without waiting for her youngsters to get down, suddenly turned tail and, upsetting the whole lot, disappeared up the trail like a whirlwind, with the cubs trying their best to overtake her.

After this last delegation had gone I waited for an hour or more, but got no more photographic opportunities. Several bears came out, but it was too dark for me to follow their actions, and none of them saw fit to run into the thread. However, just before I was about to leave, I heard something coming down the trail as if pursued by the devil, and it occurred to me that whatever it was would be in too much of a hurry to stop and examine the string, and so it proved. There was a bright flash, and for an instant the forest was lighted up, and I saw an old black bear travelling as if for dear life. I had thought that he was at his best gait before he struck the string, but in this I was mistaken. He had only been fooling along before.

Now he got down to business, and in less time than it takes to write it he was out of sight and beyond hearing. When I developed the plate it looked as though a cannon-ball of hair had been shot across it.

This was my first evening, and it did not pan out very heavily in practical results. But I had had a lot of sport, and had begun to find out, as later on I was to prove more thoroughly, that the Yellowstone Park grizzlies differ in no material respect from others of their species.

The next afternoon at about two o'clock I was again in my place of observation, with everything again in readiness for business. This time, thinking that it might not be so easy to detect, I had substituted a tiny wire for the thread. The wire was the finest that I could buy, the kind that florists use for winding flowers, and unless I knew exactly where it was, I could not see it myself when ten feet away from it. I had now selected a spot where the trail wound around between some fallen trees, where there was little danger of the bears getting scent of the wire before they came immediately upon it.

About six o'clock there came up a heavy thunder-storm and for more than an hour it rained in torrents. When I saw the storm approaching, I walked over a little way from the trail, peeled the bark from a couple of small trees, and covered my camera and my can of batteries, to keep them from getting wet. The flash-pan was fitted with a loose cover, easily thrown off by the exploding powder, and having thus protected my apparatus, I put on my rain coat, crawled under a thick-limbed, umbrellalike tree, and waited for the storm to pass. In the middle of it I saw a small black bear coming through the timber and headed for my shelter. At every flash of lightning he would make a dash for the nearest tree, but by the time he reached it the flash would be over and he would come on again. Just as he got within fifty yards of me there came a tremendous bolt, and chained lightning seemed to run down every tree. This was followed, or rather accompanied, by a splitting crash of thunder, and the small bear made one jump into the nearest tree, and never stopped till he got near the top, where he crouched down on a limb, rolled himself into a little ball, with his nose between his feet, and never moved until the disturbance was over.

When the rain had passed, I returned to where I could watch the trail, and waited for the grizzlies. It was not long before I saw an old bear coming down the trail. He was very large and fat and would, I imagine, have weighed from six to seven hundred pounds, and when I saw him advancing with the usual precautionary tactics, I was well pleased that it had rained, for I imagined that the water must have obliterated all scent, and that this old fellow was sure to run against my wire. But I was mistaken. When some six feet away he stopped, nosed his way slowly up, and stood for some seconds only a few inches from it. Then he became interested and worked a little nearer, and then there was a flash and he immediately stood up on his hind feet, much startled, and looking first in one direction and then in another. Then, like the three bears of the evening before, he started an investigation. He dropped down on all-fours, started to follow the wire toward the switch, changed his mind, worked along till he came to the little stick, and finally dug up the spool that was buried there. After thoroughly examining this he returned to the trail and followed my tracks down to where I had taken the bark off the trees. Here he nosed about for some time, and then finally turned to the right and disappeared in the timber. This negative proved to be a fairly good one, but it was not quite what I had hoped to obtain, as the bear had stopped short at the flash, while I would have preferred him in motion.

I now put in a new fuse and rearranged the camera, and it was not long before an old she bear and two cubs came down the trail; but she, after the usual preliminary examination, proved suspicious of the arrangement, and after smelling carefully along the wire, turned to the right and passed around the machine. I had brought with me on this evening a hand camera of the reflex type, built expressly for natural-history work, and I had set up my apparatus near the edge of the open park, thinking that perhaps a bear might come out in time for me to get a snap-shot of him before dark. After the old bear and cubs had passed, I crawled very cautiously to the edge of this opening and waited for them there. It was really too dark for a picture, but I thought that I might at least have the satisfaction of making a try for one. I expected that, after circling the camera, they would come out into the marsh, and this they did; but instead of passing along it as I had looked to see them do, they turned and came across it straight toward me. I was standing with the camera before my face, watching in the mirror all that was going on, and, as I remained perfectly quiet, the animals did not see me until they were within fifty or sixty feet of me. Then they went up on their hind legs, with a cub standing on either side of the old bear, and as the camera clicked, the mother dropped

From a photograph, copyright, 1906, by W. H. Wright

HE CAME ON, STOPPING AND SNIFFING

down and scuttled away up the marsh. About fifty feet from where they had stood there was a large tree, and as the old bear passed around this she was, for an instant, out of sight, and the two cubs, that had just then turned to follow her, stood perfectly still and appeared to be thoroughly mystified. Apparently she missed them about the same moment. She jumped back, and poking her head around the tree, gave two quick, short, emphatic "whoofs," and the way those cubs dropped and flew to her was a caution. She waited until they reached her side, and then gave each one of them a sharp cuff that bowled it over, and then both mother and cubs disappeared in the gathering darkness. When I developed the plate it was not even fogged by the exposure.

For another hour or more I watched my set camera. The storm had now entirely passed and the moon was shining, so that it was quite light in the little glade outside the timber. I saw four more grizzlies, including the three that had come out the night before, but they all avoided the wire. On the following evening I again tried for flashlights, and while I saw ten grizzlies, they acted in about the usual way. Not one of them set off the flash. Those that had already had experience with the apparatus did not come within a hundred yards of me, and even those that I had not seen before seemed suspicious. This night I saw an old she grizzly with four cubs, and although I have seen quite a number of black bears with that number, this was the second time that I had ever seen a grizzly with so many.

For three nights more I tried different places along the trails, but did not get another exposure. Some of them came and nosed about, but most of them turned off at quite a distance from the wire, and finally they abandoned this trail altogether, and made use of two others that ran through the timber at quite a distance from it. Finding, therefore, that the bears at the cañon had evidently taken alarm at my operations, I determined to move over to the lake, sixteen miles away, as there were also said to be many grizzlies at that point.

Here, as at the cañon, I watched the garbage pile for two or three evenings, and scoured the country thereabout during the day. Finally I decided on a trail that led out of the range of thickly timbered hills, down through some heavy woods and underbrush toward the west. Here, also, I selected a spot for my camera at the edge of a little open glade, that was covered with grass and small willows. Through this glade the main trail ran, and a branch trail also wound around at its edge near the timber. I chose the through trail for my work, because its being nearly covered with grass afforded me a longed-for opportunity to conceal the wire. I also avoided setting my camera on the ground, and fastened it to an iron spike made for that purpose, and driven into the trunk of a large tree about twenty-five feet from the path. The flash-pan was set near the same tree, and the whole effectually concealed by means of cut willow branches stuck upright in the ground. The wire from the switch was led through the long grass about a foot from the ground, and its further end tied to a small willow.

When things were thus fixed to my liking, I myself retired to a spot from which I could see some two hundred yards up the trail, and get an unobstructed view of the glade itself, and I took care to finish these arrangements early enough in the evening to give the man scent a chance to dissipate before the grizzlies came out. I found, however, that there were so many black bears in this neighborhood that I was frequently obliged to show myself in order to frighten them from the trail, and protect my apparatus from their mischievous curiosity.

The first grizzly came down the trail about sundown. He acted much as those at the cañon had, and like them, he detected the wire before he touched it. He nosed along it inquisitively, and then in a rash moment tried to claw it, when, of course, there was a flash, and he actually turned a complete somersault and disappeared up the trail at such speed that, as I discovered the next day, he fairly tore up the earth as he went.

Somewhat to my surprise my next visitors proved to be the three grizzlies that had sprung my flash at the cañon. I recognized them easily by the markings on the shoulder and neck of one of them. I may say here in passing, if it surprises any one to speak of recognizing a bear previously encountered, that there is to the full as much individuality in bears as in people, and that it is perfectly easy for me to recognize a grizzly once seen and closely examined, and under such circumstances as I am here describing I could tell a newcomer the moment he came into sight on the trail.

These three bears came up to the spot where the wire was stretched, took one good sniff, and appearing to recognize it as the same outfit with which they already had had experience, turned unconcernedly to their right and passed by on the other side.

Just before dark a she grizzly and two cubs made their appearance, but just as they arrived at the fork of the trail they stopped, stood up, sniffed anxiously at the air, and then dropped down and sidled off, with uneasy backward glances, as though they not only suspected something wrong, but feared that it might pursue them. This old bear was unusually light in color, appearing indeed, in that light, almost silvery white over her entire body, while both her cubs, from where I stood, appeared to be almost jet black. After this I waited until it was so dark that I could not see a bear in the timber, and having obtained no more shots, I returned to camp.

All this time I had been struggling against a number of difficulties, photographic and electrical. Chief among the latter was the fuse for my flash-pan. I had found no difficulty in this regard when using a shutter exposure as slow as one-quarter of a second; but if, as was apt to be the case, there was any daylight remaining, this exposure was too slow and recorded movement on the part of the animal. I had, however, succeeded in finding an extremely fine imported German-silver wire, which fused rapidly enough to allow me to use the shutter exposure of a hundredth second.

My first supply of this wire having been limited, I had ordered more, and discovered, when too late, that it was of a slightly different size; and hence, to my chagrin, when I came to develop the three exposures which I succeeded in getting, I found that my shutter had worked too rapidly for the fuse, and my plates showed no trace of an image.

At the time, however, I thought that I was getting along satisfactorily, and the next evening I again set up my camera at the same place. It now occurred to me that it might be possible, by reversing my former tactics and leaving my

scent liberally scattered over the neighborhood, to allay
the suspicions of these bears who were reputed to be accustomed to the presence of man. I therefore walked up
and down the trail for some hundreds of feet and again
concealed myself where I could watch without being seen.

The first bear delegation numbered three, but they
were not my friends of the cañon, being, for one thing,
considerably larger. I judged them to be nearly three
years old, and they would have weighed, I should say, in
the neighborhood of three hundred pounds apiece. They
were as sleek as seals, and one of them had a beautiful silver
coat. When they reached the point to which I had walked
up the trail, they stopped and scented for a few moments,
turned their heads in the direction in which I had gone,
and then came on, paying no further attention to the matter. This encouraged me, and I began to think that my
ruse was to prove successful; but when he reached the
wire the leader stopped abruptly, and the three then stood
up, looked at each other knowingly, and then, for all the
world as though they inferred a connection between my
scent and the presence of the wire, began methodically to
track me up.

I was standing near a tree, and, not having expected
any such move on their part, I had not taken the precaution to step back out of sight, and now I did not dare
to move for fear of frightening them. I therefore stood
absolutely still and watched their play with close attention and absorbed interest. They followed my every turn
as unerringly as a hound follows a hare, and came on
withal as silently as three shadows. Of course I had been
careful to select a station to leeward of the trail, and this

now helped to postpone their discovery of me. When within fifty feet of me they came to a fallen log, and, when the leader had his front paws on this, he stopped and looked ahead as though he felt that he was nearing that which he sought. The second bear started to pass him, but he turned his head and very gently took his companion by the nose with his mouth, whereupon he also stopped, and they both looked straight at me. However, as I did not move a muscle, they seemed unable to make out whether I was a living object or an inanimate one, and they again moved cautiously forward, still in absolute silence. When about twenty-five feet away, they again stood up and examined me intently, evidently doubting whether I were a bona-fide stump. Here, indeed, would have been a glorious opportunity, had I had a camera in hand, and had there been a trifle more light.

But they had come as far as they cared to. Dropping silently on to all-fours, they suddenly abandoned their investigations and bolted, only to stop at the end of a hundred feet, stand up again, again approach within fifty feet or so of me, and then turn aside and trail away through the trees. Soon after this, three more grizzlies came down to the forks of the trail. These were a trifle smaller than the others, but by far the handsomest that I had seen. Two of them were rather dark, while the third was a fine-looking animal, with a snow-white head, and silvery as far back as his shoulders. This is a typical marking, but in this case it was strangely accentuated, in spite of which, however, his companions seemed to approve of him since they had intrusted him with the leadership.

Like the others, they stopped at the string and, still like the others, they then took up my trail and that of the first three bears, and followed it as surely and as silently as the others had done. This time I took the precaution to keep behind the tree, and these three bears actually came up within ten feet of me before they discovered my presence. Then, up they went on their hind feet, and for a second there was another great picture before me: their thick, furry coats were magnificent, and the long hair standing out stiffly under their jaws lent a curious expression to their faces.

But, the second over, they too, after retreating and advancing once or twice, made their way silently into the forest.

After some waiting, an old she bear with two yearling cubs came along, apparently in a hurry, and acting as though they were late for an engagement. I thought for a moment that they were going through without stopping, but just as she reached the wire, the mother stopped short, took a hurried sniff, and then, apparently thinking it of no consequence, hurried on again. She changed her mind, however, almost instantly, but although the three turned tail and reared up on their hind legs, instead of running away, they appeared to be more curious than frightened; and it was only after rather a thorough examination of the wire and its surroundings, that they retreated up the trail, lingeringly, and with repeated glances over their shoulders.

I had hardly reset my apparatus when an old fellow came along, so huge of frame that, had he been in good condition instead of gaunt and famished looking, he would

have weighed eight hundred pounds. But he looked as
if he had not had a square meal all summer. His neck
was long, his body thin, his legs ungainly, but still showing
the tremendous muscles typical of the species. He seemed
to take four-foot strides, and I thought that surely so
great a brute would not stop to bother over a little wire;
but, on the contrary, he not only stopped, but nosed his
way carefully along until he came to the flash-pan. This
was placed upon a pole, and was about six feet above the
ground, and the old bear stood up on his hind legs and
looked into it, after which he followed the battery wires to
the camera, and then, returning, stood up a second time
and stuck his nose into the flash-pan. I am afraid that if
I had had a finger on the wire that controlled the switch,
the temptation to pull it would have been too much for me.
Meanwhile the bear, having again examined the camera,
deliberately turned back and disappeared up the trail,
and, much to my surprise, I saw no more of him.

My next visitor was the largest grizzly that I had yet
seen. He would, I should judge, have weighed close to a
thousand pounds, and he was at once so old and so fat
that it was laughable to see him walk. He was rather
dark in color, his back looked as if it were two feet broad
and perfectly flat, and his legs did not seem to be more
than a foot long. His ears were hardly visible. Perhaps
he had lost most of them in the riots and ructions of a
long-vanished youth. Hope dies hard, even at the hand
of experience, and I again flattered myself that this old
veteran would not pay any attention to my petty schemes.
He came to the forks of the trail and—took the other
turning. Then, appearing to change his mind, he turned

back and came down my trail; and, accepting this as an omen, I counted a picture of him, broadside on, as already secured. But when he reached the wire, he not only stopped and sniffed at it for several seconds, but then reared up on his hind feet, gave a snort that could have been heard for a hundred yards, and then whirled about like a demoralized coward, and tried to run. Even in my disappointment I could not help laughing at the ludicrous spectacle. I am sure that he touched the ground in at least five places at every jump, and he seemed to think that he was going along at a tremendous gait, whereas, in reality, he was making the slowest kind of time. And he was so frightened that he had no time to look where he was going; he smashed into pretty much everything he came to, and for five minutes I could hear the breaking of brush and dead branches as he crashed through the timber.

Just as I was about to pack up for the night, I heard a commotion down the trail, and looking up I saw the three bears that I had first met at the cañon coming toward me in full flight. They had evidently taken some other path to the feeding ground, and, something having frightened them, they were now coming back my way at their best gait, and quite oblivious of the wire, into which they ran with such force that it parted. This time, however, they did not stop to investigate the result, but acted as though they felt the devil himself was after them, and disappeared up the trail, at what I think must be about the limit of a grizzly's speed.

I had to leave in the morning for a trip to the Pacific Coast, and now gathered up my outfit and started for camp, supposing that my adventures were at an end.

In a knapsack on my back I had my tripod and flash-pan, the switchboard, and the connecting wires. In one hand I carried my camera, while in the other I had a covered sheet-iron box containing six dry batteries, the whole being securely tied with a small rope. I intended, on my way to camp, to see the agent of the transportation company, and arrange for a seat on an outgoing coach for the next morning; and to this end I approached the rear of a small mess house belonging to the transportation company, situated behind and on one side of the hotel. Now it happened that back of this mess house two barrels of refuse had been left standing. This was contrary to the regulations of the park, and Major Pitcher, the acting superintendent, told me afterward that all the trouble they have ever had with grizzlies arose from breaches of this rule. However, knowing nothing of this at the time, I was walking along without making any noise, and when directly back of the building, and not more than fifty feet away from it, I heard a sudden rattling among the cans, and out shot two grizzlies, followed, at a distance of about twenty feet, by a larger one. Taking it for granted that they saw me, and having, under such circumstances, no fear of the animals, I kept straight on, and thus, after a few steps, interposed myself between the last bear and the barrel from which he had been feeding. This he seemed to resent, for he turned angrily and started toward me. The whole situation developed, and indeed concluded, suddenly, so that I had no time for conscious planning. As the bear turned toward me I stopped rather mechanically, thinking that he, too, would stop before he came up to me; for I had never, in all my experience, had a bear attack me, and had always maintained that no grizzly would attack a man except under peculiar circumstances. However, this bear was either an exception to my rule, or else he considered the circumstances "peculiar."

It did not take him long to reach me, and, as he did so, he rose up and struck at my ribs with his right paw. The only weapon at hand being my can of batteries, and this, weighing about twenty pounds, being no mean defence if handled rightly, I swung it at him, hoping to stop him with the blow. As I did so, however, I advanced my left hand, and the bear's paw caught my camera, ripping out the front board and the magnet and wires attached to it. Almost at the same time I landed somewhere with the can, and, although the stroke did him no damage, it did set him back a foot or two, and turned what had doubtless been nothing but ill-temper into rage. With a loud snort he came at me again, and this time he raised himself to his full height and aimed a vicious stroke at my head; and I, seeing what was coming, ducked and closed in on him. And I was just in time, for I felt the wind from the blow, and his paw tore my hat from my head, and then, passing down the side of my face, struck me a glancing blow on the shoulder. Nothing, I think, but my nearness to the bear saved my life. Meanwhile, as I had ducked and closed in, I had swung my right hand back of me, and just as the bear delivered his blow at me, I landed mine on him; and as I had swung my can of batteries in a half-circle over my head, they came down with tremendous force.

I must have caught him somewhere about the head, for it felt as though I had struck a board, and the bear went over backward with an astonished bawl, and when he regained his feet, he was tail toward me and kept this position as long as I could see him.   The first jump he made landed him head on against a dry-goods box; at the next he smashed into a tree; but he finally got his bearings and made off, if anything, faster than he had come at me.

When I had got my bearings, I looked around for my hat, but, being unable to find it in the half light, decided to hunt it up in the morning.   I was not, at the moment, conscious of very great excitement, but when I reached camp I found that my hands were trembling rather uncomfortably, and it was several days before I recovered my usual absence of nerves.   The next day, when I rescued my hat, I found two holes in the soft brim through which the bear had driven his claws, and one corner of my iron battery case was broken open like a ship's bow after a collision.

I am quite satisfied that had I made any noise as I approached the place where the bears had been feeding, they would have retreated before I reached them; but the ground was soft and my steps were noiseless.   And, hungry as they doubtless were, one of them resented my sudden interruption of their feast.

Altogether I did not find the grizzlies of Yellowstone Park in any degree more tame or less cunning than they are to-day, for example, in the Selkirks.   Many of them, it is true, come to the garbage piles to feed, but these very bears, fifty yards back in the timber, are again as wild as any of them anywhere.   I was both surprised and interested by this, and, after watching them carefully from the positions provided for the public, I repeatedly concealed myself and watched them from the woods back of these feeding grounds. I can think of no better way to describe their actions and their attitudes than to liken them to the actions and attitudes of a man about to dive into the water. At the cañon, the garbage pile is in a hollow at the foot of rather a steep incline that leads up to the edge of the woods. Bear after bear coming down the trails that converge toward this point will stop as he reaches the brink of this declivity, glance downward, turn his head from side to side, and launch himself downhill, with the same air of committing himself to a foreign element that one sees in the upward glance and deep breath of a man launching himself from a diving-board. On their return, they invariably halted for a few seconds at the top of the hill, looked around, occasionally shook themselves, and with their first step up the familiar trail, resumed every sign of their habitual caution and alertness. While on the garbage pile itself they appear to pay scant attention to the people gathered behind the fairly distant wire fence, but even there, an eye familiar with their actions would note the constant watch they kept on what was going on and the hurried way in which they fed; and, fifty feet from the edge of the surrounding timber, they would, at the least scent or sound or sight, bolt as incontinently as in the farthest hills. Grizzlies are no more plentiful around the park to-day than they were twenty-five years ago in the Bitter Roots, and a hundred yards from the garbage pile they are no different.

XVII

FLASH-LIGHTING GRIZZLIES

IT was over two years before I had another opportunity
to tackle the problem of flash-lighting grizzlies, and in
the meantime I had brought my apparatus to the stage,
not, indeed, of perfection, but of fair reliability. I had an
improved switch, a plentiful supply of proper wire for
fuses, and two complete outfits.

Affairs so shaped themselves in 1908 that I saw my
way clear to spending the summer of that year in and
about the Yellowstone National Park, and I planned to
reach the ground early enough to be on hand soon after
the grizzlies of Mt. Washburn had come out of winter
quarters. But a serious washout on the Northern Pacific
Railroad sidetracked both me and my camp outfit and
horses, and it was well on toward the end of June when I
finally succeeded in reaching my field of operations.

I had now about given up all reliance on wires stretched
across the trails, or, indeed, upon any device looking to
inducing the bears to spring the flashes for the taking of
their own pictures, and was determined to lead the string
from the switch to my point of observation, and so operate
the apparatus myself.

This, in itself, was simple enough, and I could easily

make out to handle the business alone; but I was anxious to use two cameras, and thus not only double the chances of results by covering two trails, but make better use of the time before my presence made the animals suspicious and induced them to abandon for a time their accustomed lines of travel.   I was therefore anxious to have the help of an interested and intelligent companion.

This want, however, was not easy to supply; and as an impatient, or a careless, or even a timid, assistant in this kind of work is far worse than none, I had little real idea of trying the experiment.   I happened, though, in the interval between my two expeditions, to meet Mr. J. B. Kerfoot of New York, an amateur photographer of experience and a lover of life in the open; and as he seemed vastly interested in the casual accounts I gave him of my experiments and made some good suggestions in regard to the disposal of apparatus, etc., I finally proposed that he join me in my projected trip, and he fell in with the suggestion with enthusiasm.

At this time Mr. Kerfoot did not have so much as a bowing acquaintance with grizzlies, and his ideas about them would, I think, have pretty fairly summed up the misinformation current in the popular mind.   But these facts did not bother me, and I may take this occasion to say that he proved to be a quick student and a resourceful and most helpful ally.   He was not, however, able to join me until the latter part of July.

Meanwhile, early in that month, our party reached the cañon, and I spent the first evening of our stay watching the same trails and the same little parklike bottom where two years before I had seen so many bears.   At that time

I had remarked upon the number of females with cubs. This time I only saw one, and she had but two little ones. There was also a great falling off in yearlings and in two-year-old bears, and, all told, there did not seem to be more than eighteen grizzlies working in the neighborhood. Among these was one litter of three two-year-olds who were still running together, but most of the others were old bears.   One of these was a very large old male, nearly as broad as he was long, and that would, I should say, have weighed close to a thousand pounds.   The trails that had been continually used two years before now showed few signs of travel, but the main trail leading to and from Mt. Washburn showed that this retreat was still the home of what grizzlies remained.

This first night I set both my cameras on the same trail, and while I had arranged to spring the flash myself, I had nevertheless stretched wires across the trail in the hopes that the bears, stopping to examine them, might give me an opportunity to catch one standing on its hind feet.   I also made a different disposition of the flash-pan, and instead of placing it on top of a pole five or six feet high, I set it within a foot of the ground, hemmed in on all sides, except that next to the trail, with pieces of log and branches from trees.   I then took up my own position some fifty yards to one side in a conveniently located tangle of down timber.   Nothing of consequence, however, happened this first night, as, though several bears appeared in my neighborhood, they seemed to sense that something was wrong, and shied off before they reached me.

The next evening I replaced one camera on this trail, and set the other on a well-beaten path some distance to one side.   After some hours of waiting, the three two-year-olds came down the side trail, but stopped before reaching my wire.   Two of them then turned back, but the third apparently made up his mind to avoid the danger by circling it, and, as it happened, took the side toward the camera.   As he walked he kept his eyes upon the wire where it crossed the trail, and thus failed to notice that it extended beyond the edge of the path and still blocked his progress.   This fact he discovered too late by springing the flash, and he was then so near the camera that his feet do not show upon the plate.   The picture, however, gives an excellent idea of the formation of a grizzly's head.

While I had been watching the movements of these three bears, another old fellow had come unobserved down a trail to the left of my other camera, and then worked back toward the main trail between that camera and myself.   As the first bear set off the flash and I rose from my place of concealment, I was startled by a loud snort just behind me, and saw a huge grizzly bolt up the trail toward my second outfit.   The fright of seeing me rise from the ground just in front of him, together with the flash and the flight of the other bears, had evidently driven from his mind the remembrance of the danger that he had just avoided, and it was not until he had almost reached my second camera that he remembered for a moment where he was.   His evident dilemma was most amusing, for it was clear for a second or two that he thought of stopping, and then almost instantly changed his mind and forged full speed ahead.

I had stretched a good-sized wire across the trail, hoping that the bears, when they saw it, would stop and stand up in the course of their examination; and when the old fellow ran into this one it failed to part, but the stake to which it was fastened pulled out of the ground and, bouncing back, caught him in the side and completed his demoralization. As he disappeared in the gathering darkness he looked like a big ball of fur shot from a high-power gun. I reproduce the picture of this grizzly because it is a good example of many failures, and at the same time gives some notion of how recklessly a frightened bear dashes away, unmindful of obstructions.

The next night I tried a new trail. This time I set both cameras on it, but about fifty feet apart, hoping that the bears, frightened by one flash, would retreat and, as they often did, stand up to look back in front of the other camera. I concealed myself in a little thicket of pines, from which I had a good view of both cameras and of the trail in both directions. And I had only been here a short time when I saw the old grizzly and two cubs making their way along one of the many trails which passed near my hiding-place. They did not, however, come within range of my cameras, and I thought no more about them. It was only a few minutes, however, before I heard a snort near my thicket, and a small black bear raced by within a few feet of me and went up one of the trees among which I was standing. Peeping out to see what had so excited it I caught sight of the old grizzly and her cubs not more than fifty feet away and hot on its trail, and as I had no desire to get mixed up in any family quarrels I sidestepped and went up the next tree to the one occupied by my black friend. The grizzly, seeing this new animal appear and seem to take sides with the black bear, stood on her hind

A FLASH THAT FAILED

feet and looked at us for a few seconds, and then, knowing that we were beyond her reach, dropped down and made off through the timber.

When she was well out of the way I resumed my watch, and it was not long before I saw my friends of the other evening, the three two-year-olds, coming my way. Having already had some experience of their wariness I watched each branch of the trail with anxiety as they neared it, and drew sighs of relief as they passed these in succession. At last one of them was just in front of my first camera, and as he appeared to be suspicious of its presence, I was about to pull the string when a second one ranged alongside of him. The third was also coming on, but, being some little distance in the rear, I did not dare to risk waiting for him, and sprang the flash on the two as they stood, heads up, listening in the stillness. The instant the flash went off all three darted back and then stood up, making a beautiful picture; but they were too far away to get them with my second camera, and they ran away when I emerged from the thicket to put in a new fuse and reset my apparatus. It was now growing late and I was just thinking of giving up for the night when I saw, coming down the trail, an old bear that seemed actually to brush the trees on both sides at once. This was the old male of which I have spoken, and before the summer was over I was destined to see him many times, to spring several flashes on him, and yet never to get a perfect picture of him. At the moment, however, I knew nothing of all this, and watched in breathless anxiety as he came slowly down the trail with its many branching paths. But he passed them all, and at last, one of the largest grizzlies I have ever seen, he rounded a bend in the trail and his mighty strides brought him directly in front of my first camera. He did not stop, and I could not bring myself, for the chance of getting two shots at him, to wait until he reached the second camera. As I pulled the string, the enormous beast seemed actually to go up in the air with the flash. Then he bolted sideways, keeping his eyes on the point from which the flash had come, and, paying no attention to where he was going, he crashed into a tree with low-hanging branches, and the noise of the impact filled the forest. Then, like the rest, he stood up and looked back, and then, still with many backward glances, moved silently away.

For a couple of nights now I had no luck. The bears seemed to have taken to other trails, and, wishing to find out in what part of the woods they were working, I persuaded a young doctor who was in the party to come out with me and tend one of the cameras. I placed him and his camera on a likely looking trail and stationed myself some hundreds of yards away. I stretched the wire across the trail at his post and also ran a string from the switch to him, and I instructed him to watch in silence until a bear should stop to examine the wire, and then to whistle sharply, and, when the bear stood up at the sound, to pull the string. In the meantime I asked him to keep a sharp lookout that he might let me know later by what trails and from what directions the bears had come up.

The doctor had assured me that he was not in the least afraid of bears, and for the first two hours he was as good as his boast. I could watch him from my station. Later on, however, as the sun neared the horizon and the shadows became longer and deeper and that hushed silence settled on the woods that searches the heart of the timid in the wilderness, my friend's daylight courage began to ooze away. Any one who has never seen a grizzly in such a setting is more than likely to experience a softening of the bones when one appears, and the doctor was no exception to the rule. A couple of young bears now came hurrying down a trail a hundred feet or so to his right and, swinging his arms in the air, he gave a loud whistle, like a policeman signalling for help, and called out to me in a loud voice, "There go two."

It happened at the moment that I was watching a grizzly on my own side of the woods, and did not care to spoil my chances of a picture by going over and clubbing my friend into a more quiet state of mind, so I stuck to my post and hoped for the best. But before my bear had had a chance to get within range the doctor's whistle again rang out, and once more the information was bawled through the woods that "There go three more down the same trail!" This sent my grizzly back into the woods at double-quick and, making up my mind that that kind of help was too much of a hinderance for my taste, I took up both cameras, gave the doctor my professional opinion of him as a grizzly photographer, and went back to camp.

After the doctor's retirement another volunteer presented himself in the person of Frank, one of my camp assistants, a man who had lived much of his life in the mountains and who said he was not afraid of any bear that might show itself. So I took him along and set him to watch over the second camera, with the same instructions as those issued to the doctor, and stationed myself some fifty yards away, accompanied by a young friend who wanted to see the bear and promised to keep still. Frank was seated on a fallen tree, peering through the upturned roots at one end. Another tree, some fifteen feet behind him, lay across the one on which he was seated. He was lazily fighting mosquitoes with a small switch, and his whole attitude spoke of phlegmatic and easy attention to business.

For perhaps two hours nothing happened, and we all three remained at our posts. Then the young man at my side whispered to me that there was a large brown bear some twenty-five yards in our rear, and as I turned to look, I saw the huge grizzly that I had taken a snap at a few nights before sneaking up behind Frank with extended nose and every appearance of puzzled curiosity. Frank was still gazing between the roots at his camera and lazily swaying his switch, and, knowing that the big grizzly was aware of his presence and only trying to satisfy his curiosity, I made no sign. When the bear reached the dead tree that lay at right angles to Frank's seat, he placed his fore-paws on it, stretched his head out, and began to sniff—well, out loud.

Frank turned around casually at the sound, and for a hundredth of a second there was a tableau that I would have given a good deal for a picture of. Then Frank's switch began a frantic tattoo on the nearest root. Frank himself leaped to his feet and the fat old grizzly shot away sideways—as luck would have it—in our direction. His first two or three jumps covered more than half the distance between us, and, as it began to look as though we would be trampled, we also jumped up, when, with a loud snort, the old fellow again changed his direction and made off puffing and blowing. Frank allowed after that that, although he was not afraid of bears, he would just as soon not have anything to do with any animal of that size that could get up near enough to smell the back of his neck without his hearing them come. So for the few remaining nights of our stay I went it alone.

However, the bears were by now becoming very crafty, and it was so difficult to tell to what unused trail they were resorting that I got but few shots. On one occasion, however, I secured a picture which, though not in perfect focus, illustrates very clearly one of the characteristics of these animals. Two good-sized grizzlies were coming down a winding trail on which I had posted myself, and the larger of the two, happening to notice my apparatus before he reached it, turned out of the trail and approached the camera with a curiosity that looked sufficiently like ferociousness to disturb any one not familiar with the creature's habits. Finally, just as he paused with one foot raised and his nose extended, I sprang the flash, and he almost broke his neck in his haste to get away. The next day we moved camp to the shores of Yellowstone Lake.

Here, in order to be nearer my field of operations, I separated temporarily from the rest of the party and made a small camp near the bear ranges, and for two nights worked on my old grounds of two years before. Here, too, the number of bears had greatly decreased, and the only animals of any marked interest seemed to be a mother and three cubs. On my return to camp the second evening I found Kerfoot waiting for me, and the next day I took him with me for a general survey of the ground.

We determined, after a pretty thorough discussion of the matter, to devote ourselves to securing a picture of this grizzly mother and her family; and as, after my experience with my volunteer helpers the week before, I did not propose to try any more amateurs without having them serve an apprenticeship, I only took one camera out the next night. We set that on the trail the she grizzly had been using, and stretched a fish-line to our hiding-place about a hundred feet away. We placed no wire across the trail, and from this time forward never again used that device.

The old lady and her youngsters proved to be on time, and just about seven o'clock I caught sight of them in an open glade to our left. It was my first near view of her, and I saw at once that she was as cross-grained and ugly a brute as one was likely to meet in a lifetime. I instantly decided that two hundred feet was none too much space to interpose between her and the man who proposed to challenge her with a flash, and, whispering to Kerfoot to come quietly, we backed away, unrolling the fish-line as we went, until we came to the end of it. The mother was an old bald-face; one of the cubs was almost black, one was gray, and the third had a silvery sheen to its fur. The whole family came on, the mother looking neither to the right nor left and paying little attention to the cubs that ran, now ahead and now behind her. She did not seem to suspect the presence of the camera, and when she and two of the cubs were opposite the lens I pulled the string. The flash, however, refused to explode, and when I pulled again, to my utter disgust the string broke. It proved to have become entangled with an intervening branch, probably during our retreat, and the incident, costly as it was, taught us a lesson that we never forgot. Indeed, from this time on we almost invariably handled our cameras from the branches of some convenient tree, thus guarding against any necessity of changing our positions unexpectedly, and we took great care and considerable pains in so running our strings that they were sure to work smoothly when needed. We got no other shot this night, and returned to camp, feeling that we had lost the chance of our lives.

The next night we again went after this old bear, but she did not put in an appearance; and the third night, feeling that Kerfoot was now able to handle that part of it himself, I left him in charge of the flash-light apparatus, and stationed myself, with a natural-history camera, in a tree by the little open glade where we had first seen the old bear and her cubs.

About seven o'clock she made her appearance, but, instead of going on down the trail toward where Kerfoot was waiting for her, she and the cubs stopped to dig roots at the side of a small marsh. I now understood why we had not seen her the night before, and while she offered me several opportunities for long-distance shots with my hand camera, I refrained from risking them in hopes that she would ultimately move on in the right direction.

It was just beginning to get dark when I saw a medium-sized bear come up the trail from the direction of the flash camera, and when the she grizzly saw him she called her cubs and retreated into the woods. I was at a loss to explain this, as she was larger than he, but the sequel led me to believe that she had mistaken him for a much larger bear occasionally seen in the neighborhood. At any rate, she reappeared in a few minutes, took a good look at the bear that was now sampling the roots that she had been digging up, and then, with a snort, took after him and chased him clear out of sight. She then returned, grunted to the cubs, and stalked off down the trail, looking as though she was mad as a hornet.

She had hardly disappeared from sight when the whole woods was lighted up by a flash, and this was followed by a roar that fairly made the trees shake. About ten seconds later the roar was repeated, and when I thought of my tenderfoot apprentice perched in his tree, I could not help laughing over the initiation he was getting in the art of photographing grizzlies.

A little later on, when I went to join him, I found him packing up the apparatus, but casting nervous glances over his shoulders the while, and the next day, when we developed the plate, I told him that I guessed he'd do. We worked two more nights in these hills, but got no other shots at any bears.

We now spent some weeks camping on Shoshone Lake, in the Snake River country near Jackson's Hole, and along the Firehole River, and while in this neighborhood spent three evenings in the hills near the lower geyser basin. We had found few signs of grizzlies in this place, although black bears seemed to be plentiful; but one trail, that led straight up the wooded hillside to the plateau six or eight hundred feet above the level of the basin, was evidently used by them.

We followed this trail for some miles, hoping to find the daylight retreats of the grizzlies that used it; but

From a photograph, copyright, 1900, by J. B. Kerfoot

SHE WAS AN OLD BALD-FACE WITH THREE CUBS

when we turned back it was still stretching away like a boulevard in the direction of the lake.

We had seen a group of four really magnificent grizzlies, apparently three-year-olds, come out of these woods the first evening of our stop, and were more than anxious to get a picture of them; but though we spent two long evenings beside the big trail we never laid eyes on them again. The first of these evenings it was raining hard, and we had covered our cameras and had turned down the hinged tops of our flash-pans, and were huddled under such shelter as we could find, when a red squirrel ran up the tree to which our switch was nailed and sprang the flash. There was an explosion that was heard two miles away at our camp, and we were at a loss to determine whether it was caused by the hinge of the flash-pan top having rusted, and so confined the charge, or whether some new powder we were using for the first time was unreliable. The next week we learned that it was the latter.

On the last day of August, having left the other members of the party at Gardner, Mr. Kerfoot, Frank, and myself started back for a week's final photographing in the neighborhood of the cañon. There had, meanwhile, been a heavy storm; the mountains were white with snow, and the air was clear and wintry. We reached our destination and made camp by two o'clock on the first of September, and Kerfoot and I climbed to the scenes of my earlier efforts and made a hasty survey of the region now so familiar to me. It was already late, and, as we had no time to seek out new locations, I placed my camera near the point where I had been treed by the old bear and cubs, and where I had secured some of my earlier pictures, while Kerfoot placed his apparatus at the intersection of two more travelled trails.

We found that while the same bears were to be met with as earlier in the season, they were, while fatter and sleeker of coat, more cautious than before, and it was all but impossible to so place and hide our apparatus that they would not detect it.

This first night I secured the two pictures that I have called "Grizzlies Feeding" and "A Grizzly Walking Out of the Woods"; but it turned out that Kerfoot had had a disheartening time of it, and, while many bears had come his way, a change in the direction of the wind about sunset had wafted the scent of his apparatus up the trail and warned the approaching animals of his presence. And for several nights thereafter we had little success.

Kerfoot by this time had become both enthusiastic and expert, and, as the weather was fine, the moon at the full, and our last days slipping away, we stayed late in our trees and did not give up till the bears themselves retired. We sometimes worked as much as a mile apart and, to the best of our skill and ability, covered the whole range. But though we had to make our way home through the dense forest that we knew contained many large grizzlies, none offered to molest us as long as we were moving openly and with some noise through the woods.

One evening, however, I had set my camera beside a well-worn trail that ran along the marshy bottom of a valley, and I had led the string to a big tree well up the steep hillside, whose three trunks offered me an ideal screen and peep-hole. I was working on the ground this night, and just back of me and my tree there ran a faint and little-used trail. I had been on watch for some hours and had not seen a sign of any bears when, hearing the sharp panting sound that, unlike the black bear, the grizzly always makes when excited and running, I turned sharply round and saw a bear coming full tilt up the disused trail and within a few feet of me. I instantly dodged around my tree, broke a dead limb to make a noise, and let out a "whoof!" to scare the bear. Then, as he passed, I stepped out again and almost faced a second grizzly that was following him. I dodged again, and as this one passed me he raised one paw as if to strike, but he did not pause, and was soon out of sight.

What had startled them I do not know, but they were evidently bent on getting away from something and did not propose to have their retreat cut off. As for myself, I felt that I had had enough grizzlies for one night, and pulled out for camp as soon as I could get my things packed up.

The next night, while working in quite another part of the hills, I got a splendid picture of these two bears, fine fellows in the pink of condition, and this time they showed no desire to resent my presence, even standing up at a distance of fifty yards or so and looking at me while I reset my camera.

But if the park bears, like all others I have ever known, showed no disposition to molest us unless interfered with, there were individuals among them that paid no respect to our camp. Early in the summer we had been much put to it to protect our cook tent from them, and on one occasion one of the men had even driven the point of a prospector's hammer through the tent and into a bear's nose. And on this second visit we fared even worse, until we hit upon the expedient of setting our flash-pans just outside the canvas walls at night, and baiting them so that thieving bears would set off a small pinch of powder in their own faces. One night four grizzlies visited us in succession, but by this device we were spared a repetition of our first night's experience, when our camp arrangements were literally wrecked and the inside of our tent in the morning looked like a hurrah's nest.

One of our keenest desires was to get a good picture of the big bear that I had already taken once, and that had frightened Frank the night he went out to help me. Kerfoot had got one shot at him, but the powder burned so badly that the bear had time to move his head clear across the plate. This animal, the undisputed lord of the range, was always the last one to appear, and now that it was drawing toward fall it was very dark when he came out. The night after I had photographed the two bears last mentioned we had determined to place both cameras upon the main trail from Mt. Washburn, hoping to get a shot at this old fellow, and determined, if necessary, to spend the night in our trees. I was stationed near the foot-hills, and Kerfoot about a mile farther down, and we hoped if one missed the bear that the other would be more fortunate. We had contrived our seats in the afternoon, and mine was high among the branches of a clump of firs that screened me completely from the trail, but left me a small hole through which to keep watch.

For several hours nothing appeared, and finally it became all but impossible to see except where the moon cast a stray beam of light among the shadows. Then, at last, in one of these light spots, I saw a dark shape that could only be the big bear I wanted. If he passed the next lighted space and kept on into the farther shadow he would be in range of my camera. So, to be ready for him, I reached for the string that I could no longer see, and I suppose in doing so must have made some slight sound. At any rate, the bear did not enter the second lighted space, and a few moments later I saw him pass across a moonlit park, three hundred feet or so to the left of my camera. Slight as was my movement, he had either heard or seen it, and veered off.

And a mile down the trail Kerfoot, a little later, had the same experience. He caught sight of the big fellow just as he reached the camera, and in reaching for the string touched a branch, and the big bear shied instantly and circled out of range. The same night, and just before the big bear came along, Kerfoot, by guessing at the position of the animals that he could not see in the darkness, sprang a flash on two bears, whose picture I have called "Three-Year-Olds."

It was not till the last night of all that I got another chance at the big fellow. Early on this evening I had flashed a pair of young bears, and the powder, which had been acting badly throughout our stay, again exploded instead of burning, and bent my flash-pan quite out of shape. I had repaired the damage as best I could, and had reset my apparatus when, about nine o'clock, my big friend came along. He was just passing a small fir-tree when I pulled the string, and it seemed to me, and I dare say to him, as though the end of the world had come. The flash exploded with a noise like a twelve-inch gun, and a shower of burning particles rose in the air and glowed for more than a second.

When we developed the plate we found that the old bear had backed up against the tree and, with bared teeth and savage mien, had faced the unexpected danger. The picture was far from perfect, but it gives a notion of his splendid proportions and of his savage courage. I have called it "At Bay."

And so ended our summer's work. We brought away a thousand memories and about a dozen flash-light photographs.

PART III

CHARACTER AND HABITS OF THE GRIZZLY

XVIII

DESCRIPTION AND DISTRIBUTION

A S we have seen in the classification of bears, science recognizes, besides the Rocky Mountain grizzly (*Ursus Horribilis* Ord) and the very distinct Barren Ground grizzly (*Ursus Richardsoni* Mayne Reid), two sub-species of the Rocky Mountain type—the Sonora grizzly (*Ursus Horribilis Horriæus* Baird) and the Norton Sound grizzly.

Taken together, these animals range from the mountains in southern Mexico, throughout the Sierras and the Rockies, all the way to the Barren Grounds; but *Ursus Horribilis* Ord, the true grizzly bear in the sense that it was the species first recognized by science, has a more wide-spread geographical distribution than any of the others. While its type locality is, of course, Montana, where it was first discovered, its proper range extends from Wyoming and northern Utah, throughout the Rocky Mountain chain, to Norton Sound, Alaska; and neither in Montana, Idaho, Wyoming, nor in the interior of British Columbia south of the Barren Grounds, is any other species to be encountered. On the other hand *Horribilis*, Ord wanders, to a considerable extent, into the domains of other grizzlies.

I wish once more to call attention to the fact that, throughout the following chapters, the term "grizzly bear," when unqualified, refers to the Rocky Mountain grizzly, *Ursus Horribilis* Ord.

This bear is generally considered by scientists to be somewhat smaller than the Sonora grizzly, also smaller than the Norton Sound grizzly, but rather larger than the Barren Ground grizzly. He is described as having hairs elongated over the shoulders so as to give almost the effect of a hump; and this collar, or "roach," while typical of the species, varies greatly in development in individuals, so that one sometimes sees a grizzly with no noticeable ruff, and again sees specimens with this feature so marked that the old hunters claim "the roach-back" to be a distinct variety.

The grizzly is also described as having larger fore claws than any other species, and these of a whitish color and nearly straight. This last distinction will not, how-ever, judging from my own observations, apply to the species as a whole. I have seen many hundreds of these animals, and, while the claws of some are white, as de-scribed, I should say that a majority of them have dark nails, and these rather curved than otherwise. Of course, as to this, it makes a great difference whether the bear is examined in the early spring when his claws are fresh-grown, or in the late fall when they are worn with much use. The accompanying illustration will give a better idea than any description. These powerful claws are, when not worn down by use, from four to six inches in length; they are narrower at the base in comparison to their length than those of other bears, and do not curve sharply

1. FRONT FOOT OF A BLACK BEAR
2. FRONT TRACK OF A BLACK BEAR
   Size, 5 x 4 inches.

3. FRONT FOOT OF A GRIZZLY BEAR
4. FRONT TRACK OF A GRIZZLY BEAR
   Size, 8 x 4½ inches.

downward and backward as do those of the black bear;
nor are they so sharply pointed. But, as may be seen from
the photograph, they cannot be justly spoken of as
"nearly straight."

The hind paws of all bears differ from their front paws
more radically than those of most quadrupeds, and while
those of the grizzly do not differ from those of other bears
as markedly as do their fore paws, they are none the less
easily distinguished. The tracks left in snow or mud or
dust by these hind paws bear an uncanny resemblance to
the mark of a human foot, and it is to this the animal owes
its nicknames of "Old Ephraim" and "Moccasin Joe."

It may, perhaps, be of interest to note, by means of
reference to the accompanying photographs, the more
salient and easily recognized differences in the tracks of
the black and the grizzly bear, these animals being frequently met with in the same regions. On the fore paw
of the black bear the pad is quite round at the front, and
slightly convex at the rear. It does not have the indentation found on the inside of the grizzly's front foot,
and is, roughly speaking, somewhat kidney-shaped. The
outer side of the pad is also much narrower than that of
the grizzly. The claws are not so long and are different
in shape, curving much more at or near the point, and
they are proportionately broader at the base, growing
more like the claws of the cat. The foot is also much
thinner. Both the front and hind feet have greater muscles, presumably developed by climbing. The black bear
does not do so much digging as the grizzly, and therefore
does not wear his claws down so short in proportion to
their length as does the latter. To be sure, he turns over

logs and rocks much as the grizzly does, but he uses his entire paw for this purpose, whereas the grizzly will often use a single claw, almost as though it were a finger. The black bear is an agile climber, and, at times, when forced to climb against his will, will cling to the trunk of a tree and circle it like a squirrel to keep out of sight of the hunter.

On the grizzly's front paw the pad is only slightly rounded out in front, while in the rear it is convex instead of concave, as in the case of the black bear. It also shows a decided indentation just back of the big toe, and the marks left on the ground by the long curved claws are at a distance of two or three inches from the ends of the toes. These claws are not at all adapted to tree climbing, as will be seen at a glance, but are used for digging roots, turning over logs and rocks, and are invaluable for fighting. These claws vary in color; sometimes they are striped, sometimes white, and sometimes almost black. The three middle ones are much longer than the two outside ones.

There are also distinct differences in the hind feet of these two animals. One salient characteristic of the grizzly's hind foot is a protuberance on the outside of the pad just back of the little toe. This is not found on the hind foot of the black bear; but the latter has, on the other hand, two small protuberances on the inside of the pad, one near the big toe and one near the heel. In some cases these are more noticeable than in others, but their impressions are always to be found. Another noticeable difference lies in the shape of the heels. That of the grizzly is sharply pointed, and if a line be drawn from the middle toe along the axis of the foot, the point of the heel

1. HIND FOOT OF A BLACK BEAR 3. HIND FOOT OF A GRIZZLY BEAR
2. HIND TRACK OF A BLACK BEAR 4. HIND TRACK OF A GRIZZLY BEAR
Size, 8 x 4 inches. Size, 10 x 5½ inches.

will be found to lie to one side of it. The heel of the black bear is blunt and rounded, and a line from the middle of the toe along the axis of the foot will exactly strike the point of the heel. Again, in the grizzly's hind foot the front line of the pad next to the toes is but slightly curved, while in the case of the black bear it is rather strikingly rounded out. The hind foot of the grizzly is also much slenderer than that of the black bear.

In comparison with the black bear, the grizzlies are longer of body and straighter along the back, and do not show the same marked hump over the haunches. They are narrower in the forehead, their jaws are longer, and their muzzles much less tapering. This latter characteristic is very marked, and appears even in old bears that have grown fat. Several of the flash-light pictures reproduced in this volume show this square muzzle excellently. The grizzly's fore legs also differ greatly in build from those of the black bear, being smaller in the ankle and marked by a heavy and symmetrical development of muscle.

Another peculiarity of the species, also found in varying degrees in the allied group of the Alaskan brown bears, is the great length of the third incisor tooth on each side of the upper jaw. This almost gives the grizzly the appearance of having four upper canines, and the lower canines, fitting in between these upper teeth, give him a peculiarly formidable armament.

The color of the grizzly bear has given rise to much discussion. This is no doubt partly due to the implication in the common name of the animal, and partly to the unusual and apparently endless variability of the species in this respect.

That Lewis and Clark found striking differences in color among these bears is, as we have seen, amply attested by their journals. They not only refer to them in varying terms as the "grizzly," "gray," "white," "brown," and "variegated" bear, but put down at some length their own conclusions and the opinions of the Indians in regard to the bearing of these color variations upon species. Nor in this respect is there any difference between their day and ours.

I have seen grizzlies in all shades of color, ranging from almost jet black, through the browns and creams, to practically snow white; and it is, indeed, rarely that one finds two of exactly the same color. It may be well, at this point, to say a word in regard to the use of the word "white" with reference to the grizzly bear. The word is not used in a spectroscopic sense, but exactly as we use it when we say that a man has white hair. If it be permissible to say of an old man that "he had snow-white hair," then I have seen grizzlies of which it was permissible to say the same. I have seen them as white as a mountain goat, or as white as what is commonly called a "goat-skin" rug. It is not uncommon to see an old she bear with three cubs, each of a different color; one, for instance, of a dark brown, verging almost upon black, a second of a light buff, and the third nearly white, or white as far back as the shoulders. In fact, we seldom see a litter of cubs that are all of the same color as the dam, or, for that matter, all of the same color. Whence comes this great variation of coloring I am wholly unable to say. It is, however, an indisputable fact and typical of the species throughout its range.

It is interesting to note, however, in this connection,

that a great variability in color is also shown by the black bears throughout the Rocky Mountain region, where these animals are to be found of various shades from dark brown to buff, and even, occasionally, of a uniform cream color. I have never seen a jet black grizzly, or a black bear as nearly white as some grizzlies; but between these extremes I have seen many bears whose species was not to be determined by color alone. Indeed, throughout the North-west, so far as the color of either the grizzly or the black bear is concerned, it no more determines the animal's classification than does the track indicate its color. One often hears the cinnamon bear spoken of as a distinct species, and this idea is, I think, widely entertained; but while there are many bears of a cinnamon color, they may be either of the "grizzly" or of the "black" variety. I have seen cinnamon colored bears in both species.

One is often asked as to one's opinion as regards the "typical" color of the grizzly bear. This is a natural question, both from the stand-point of the layman and from that of the naturalist; for it is customary, in describing an animal for the purposes of scientific record, to select a typical specimen, and base the description upon that. But having myself seen so many hundreds of these animals, and having seldom seen any two exactly alike, I have always both hesitated to give a categorical answer to this query, and questioned in my own mind the value of an arbitrary selection. Yet, in view of the emphasis often laid upon the point, I wrote to my friend, Dr. William T. Hornaday, of the New York Zoölogical Society, and asked him what his observations have been.

Dr. Hornaday says: " Regarding the color of the grizzly

it remains to-day just as Lewis and Clark found it a century ago. I have never taken the pains to acquire from other authorities any special information about the color of the grizzly bears—chiefly for the reason that I have myself seen a sufficient number of skins, and also living bears, ranging all the way from Chihuahua, Mexico, to White Horse, Yukon Territory. I have seen, all told, at least two hundred skins of grizzlies, and not far from forty living bears. Having seen these variations for myself, I have never taken the pains to collect data from books, but have learned much from practical, bona-fide bear hunters like yourself. So far as I can judge, the color of the grizzly conforms with no known law of coloration. I do not know of any other bear species in which the coloration of the pelage is so erratic as it is in that of the Rocky Mountain grizzly. At this moment we have* a female grizzly from Colorado which is very dark, and so nearly destitute of the usual light color on her hair tips that she is at times not easily distinguished from the Alaskan brown bears—the latter being wholly without grizzled hair tips. With this bear is exhibited a medium light-colored grizzly from Wyoming, which, to judge by color alone, might well be called another species. It is the kind frequently spoken of as the bald-face grizzly—the hair of its entire head being of a light buff color.

"I have seen grizzly bear skins from El Paso, Texas, said to have come from old Mexico, that were almost a golden yellow. Our grizzly bear from White Horse, Yukon Territory, is about the same color as the so-called bald-face from Wyoming. The hair of some grizzlies is

---

* In the Zoölogical Park, New York, of which Dr. Hornaday is Director.

light at the base, and in others it is dark.   In my opinion, there is no way in which to explain the erratic color of the grizzlies.

"Unfortunately, there is no type specimen to which naturalists can refer.   The animal was described by Lewis and Clark in varying terms, according to the different colored specimens that they saw; but they brought back no skins for permanent preservation, as representing the species.   Mr. Ord merely describes the animal on the strength of the writing of others, and not according to what he himself saw, and he had no type specimen.   It is very unfortunate that this important species should have been described without a type specimen.   Inasmuch, however, as it *was* so described, it remains for you, or any other author, to describe according to his own observations what should be regarded as the standard and the most *typical* color.   For myself, were I to choose a skin to offer to the world as a type, I would pick one which is dark colored, but having the terminal third of its longest hairs colored *gray*.   In other words, the grizzly-gray wash on the coat *should not be ignored*.   The very light and the very dark phases I should regard as extremes, and in no sense typical."

Such a skin as Dr. Hornaday describes does, undoubtedly, represent that non-existent thing, the "average grizzly."   That is to say, it is, more nearly than any other, typical of the abstract idea of this bear entertained, not only by the general public, but by those who have seen a good many specimens.   But as a matter of actual observation I myself would not select this kind of hide for a typical one.   My experience has shown me that there are, perhaps, fewer bears of this particular color than of any other, and if one is to be governed by the comparative commonness of a particular type, I would select as a typical hide one with the head very light in color and shading away to a point back of the shoulders, and the remainder of the body rather dark, with the long hairs slightly tipped with white. It is to be noted that, among the bears described by Dr. Hornaday as at that time in the Bronx Zoo, the one from Wyoming and the one from Alaska approximated to this description. I have seen more grizzlies of this general color scheme than of any other, and that, too, in all parts of their range.

Of course the time of year at which the grizzly is seen may, and, I believe, often does, make a difference in its color; for a bear that is dark in October may be a rusty-red or a golden-brown in June. But, so far as I am able to judge, this change in the color of the pelage of a particular bear occurs only in those of one dark color, and, even then, is the result of particular conditions, and follows no established or dependable rule. A bear that is white about the head and shoulders is, I believe, more than likely to remain this color at all times.

At one place in British Columbia, where I hunted for a number of years, there was a bear that we saw year after year that was apparently snow white from the tip of his nose to the end of his tail. This bear was seen not only by myself, and by various gentlemen who accompanied me at different times, but by others at different times of the year. I have also seen and recognized, year after year, grizzlies of a dark hue, and know for a certainty that they changed scarcely at all in color.

It is no doubt due to this remarkable range of coloring that hunters and trappers have so persistently maintained that there are many different kinds of grizzlies throughout the Rocky Mountains; such as the "bald-face," the "roach-back," the "silver-tip," the "range bear," and now and again the *real* grizzly. These old fellows, good hunters, no doubt, but infrequently discriminating observers, have insisted that there is a difference as to species in these different colored bears, and that some are more fierce than others; that a roach-back or a bald-face, for example, will give the trail to neither man nor beast. But this, in my experience, is all a myth. I have never found that those of one color were any more ferocious than those of another; or, indeed, that those of one color differed from those of another color in anything *but* color.

The size of the grizzly is another mooted question. There seems to be a wide-spread idea that a grizzly must weigh a ton or more. Even in the mountains, among those who ought to know better, I have often heard the statement that "it could not have been a grizzly as it was not large enough." Now, to whatever size the grizzly may attain, it is not born weighing a ton. As a matter of fact, as we shall see later on, they weigh only a few ounces at birth, and, so far as my observations go, I am inclined to think that they are rather smaller at birth than the black bear. I have often seen females of both species, with their cubs, in July, and have found that the grizzly cubs were not nearly so large as those of the black bear; although the mother grizzly was fully as large again as the black bear dam.

Moreover, full-grown grizzlies differ vastly in size. I have myself killed small grizzlies, and have seen such killed by others, that were extremely old. Their teeth were worn down to the gums, and many of them decayed and broken off, and there was every other indication that the animals were full of years; and yet these bears would not have tipped the scales at more than two hundred and fifty or three hundred pounds. Again, the finest and largest specimens of the grizzly I have ever seen appeared to be in the prime of life. Their teeth were white as ivory, and they were the personification of strength, size, and vigorous bearhood.

Unfortunately, the weight of the grizzly has been more a matter of guess work than of knowledge, as few of them have been reliably weighed. Some that have been raised in captivity have, at their death, been carefully weighed; but none of these have tipped the scales at any such weight as some old hunters have claimed for the animal.

Dr. Hornaday says, regarding the weights of grizzlies raised in captivity: "The largest grizzly of which I ever have had an authentic record is the one which lived and died in the Lincoln Park Menagerie, Chicago, and which was weighed to oblige Mr. G. O. Shields. Its weight was eleven hundred and fifty-three pounds. That is, I am certain, the heaviest record for any grizzly whose weight was ever ascertained by scales."

The largest grizzly that Lewis and Clark killed, or at least the largest they mention, they "conjectured" would weigh a thousand pounds. This bear measured, as they state, nine feet from the end of his nose to the root of his tail, and a bear of these dimensions is a very large animal. Judging from my own observations, I should say that they underestimated, rather than overestimated, its weight.

From a photograph, copyright, 1909, by J. B. Kerfoot

THE GRIZZLY'S WALK IS A KIND OF SHUFFLE

James Capen Adams, according to Mr. Hittell, claimed that his grizzly bear Sampson had been weighed on a hay scale, and tipped the beam at fifteen hundred pounds.

I have never weighed one myself, nor have I ever seen one weighed. Many years ago there was a large grizzly killed in Idaho, hauled into Spokane, and sold to a butcher, who claimed that he weighed it and paid for eleven hundred and seventy-three pounds of bear meat. These figures were placarded on the carcass as it hung in front of the shop. I, with many others, saw the bear, and, had it been recently, I should certainly have had it weighed and had the weight certified to. But at that time no one thought of doing such a thing, and thus, unfortunately, the statement of the weight of one of the largest bears I have ever seen is uncorroborated.

The gait of the grizzly is peculiar and almost impossible to describe. His walk is a kind of shuffle, similar to, yet very different from, that of the black bear. It is less awkward, less loose-jointed, and yet in a way more ungainly, because more suggestive of power. It differs, moreover, from that of the black bear in that at no time while in motion does the grizzly give one the impression of being off its guard, or, so to say, of there being any point in its stride when its muscles are not under instant and complete command. Any one familiar with these two bears can tell from one glance at the back or at the legs of an animal in motion to which species it belongs. I have mentioned a case in which I mistook a grizzly for a black bear, and did not discover my error for some time. This, however, was wholly due to the fact that my original assumption closed my eyes, and, having once definitely accepted the animal as a black bear, I simply let it go at that. The grizzly's walk has the appearance of leisure, but he will cover long distances in a remarkably short time without once going out of it. His run is something between a lope and a plunging gallop. He can outrun the black bear by nearly a half, no man can match him in speed, and it takes a mighty good horse to catch him.

XIX

CHARACTERISTICS AND HABITS

GRIZZLY cubs (it seems, on the whole, just as well to begin at the beginning) are brought forth in the winter den, and at about the same time as those of the black bear; that is, about the end of February or early in March. The period of gestation is seven and one-half months.

Ordinarily a grizzly has either two or three cubs to a litter, with the chances somewhat in favor of the smaller number. Twice I have seen an old grizzly with four cubs, and, very occasionally, one with only one. In the summer of 1906, in one locality, in the mountains on the eastern side of Yellowstone National Park, I saw two old grizzlies with two cubs each, two with three each, and one with four.

After the cubs are born, the family continues to occupy their den for a month or two, according to the locality; and during this time the conditions make it impossible for the mother to get food. Whether she drinks or not I am unable to say, but she could get water only in the form of the snow that has closed them in. As no food is laid up in the den, and as the old bear does not come out to forage, we know for certain that she does not eat during this time.

We also know that the cubs must be at least two months old when they emerge from the den.

When the cubs are first born they are wee little chaps, not larger than the common gray squirrel, and I do not think that at the time of birth the size of the mother makes any material difference as to the size of the cubs; although I have reason to believe that, as is often the case with other animals, when but a single cub is born to a litter, it is likely to be larger than where there are two or more. I have examined the little fellows shortly after birth, and where the mothers differed greatly in size, but the cubs were always of about the same weight.

Adams, in describing the capture of Ben Franklin, refers to the enormous size of the old mother and to the smallness of the cubs, whose eyes were not yet open. He says that he carried them both in his shirt front, and it will be remembered that he put them to nurse to a greyhound that happened to have puppies of about the same age, and he says that the cubs were little, if any, larger than the puppies whose places they usurped.

As I had never actually weighed a young grizzly cub, and as I had found that many persons unacquainted with the facts in the case seemed to regard it as impossible that the young of so large an animal should be so small, I asked Dr. Hornaday whether he had ever weighed any cubs born in captivity. He replied that "a grizzly cub, which was born to our Colorado grizzly, 'Lady Washington,' on January 13, 1906, weighed eighteen ounces."

A year or so later, on the evening of January 18, 1908, while in New York City, I received word from Dr. Hornaday that two cubs had just been born to a Rocky

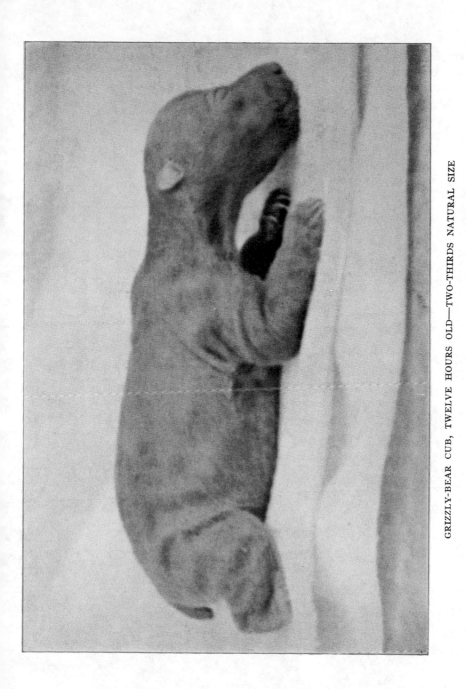

GRIZZLY-BEAR CUB, TWELVE HOURS OLD—TWO-THIRDS NATURAL SIZE

Mountain grizzly in the Bronx Zoo, and that if I would come up the next morning I might see them, and photograph them. They were just twelve hours old when I arrived, and the two weighed one pound, that is, eight ounces each. They measured nine inches from nose to tail, and the accompanying illustration gives an excellent idea of their appearance. The mother weighed about six hundred pounds.

Grizzly cubs are born with their eyes closed, and do not open them for about the same period as puppies and kittens. There is very little hair on their bodies at birth, and what there is is so short that they have every appearance of being naked. They are born without teeth, or, rather, their teeth are so little developed that they can barely be felt by pressing one's finger down on their gums. These teeth, however, grow very rapidly, and are early replaced by a new set as sharp as needles.

The dam and her family leave the den anywhere between the first of April and the middle of May, according to the locality. I have never found the fresh track of a grizzly in the Kootenai country earlier than the 5th of May, while in the southern part of central Idaho I have seen where a grizzly had left his den as early as the middle of March. The male bears leave their dens from one to three weeks before the female and her cubs come out; yet in any one locality, nearly all of each class leave their dens about the same time.

How the brutes can tell just when to come out is one of their own secrets. In the Selkirks they den so high up among the peaks that when they emerge there are from four to six (and in some cases even ten) feet of snow still lying over the country like a great white blanket. Only on the slides, which have been swept by the tremendous avalanches that usually come down in March, is the ground clear. Yet on one of our trips to this region we saw where thirteen grizzlies came down the mountain side in a single night. They all came down an open place not over half a mile across, and it was in following their trails back up the mountain that I found the six dens hereafter mentioned. These were all natural caves among the cliffs, their mouths well concealed by thick firs and juniper brush, and the animals, in coming out, had broken through some five feet of snow. As it is, therefore, not the melting snow that arouses them, it would seem that there must be some kind of nature's alarm clock, known to the bears, that informs them when it is time to get up.

It is some months before the young cubs begin to forage for themselves, even in part. Dr. Hornaday, speaking of bears in captivity, writes me: "I think the average age at which a grizzly cub begins to feed independently of his mother's milk is about four months. Of course, the beginning on solid food is made very slowly, and the youngsters nurse vigorously all summer."

I am, however, sure that this weaning process begins later in the wild state. I have many times seen a mother grizzly digging roots and feeding on grasses in August while her cubs were running about or lying in the sun, and seemed to take no interest whatever in the food that she was so busy in getting. Yet a bear cub, when the time comes, knows just as much about the proper food of the bears in the locality where it was born as does the oldest bear on the range.

I have caught them when they first came from the den, and when the earth was covered with a thick mantle of snow, and have then taken them from their mothers, and for weeks fed them milk and such soft foods as I could make them eat; and then found that, when they were old enough, they would go out and, wholly untaught, select the food that the old bears were eating at the time.   One black bear cub that I took in this way I kept with me all through the summer and fall hunting, having caught him early in June, in the Bitter Root Mountains, when the ground was covered with snow and when he first came from the den.   He was then a little fellow not larger than a common house cat, and cried for his mother the greater part of the time for a couple of weeks.

This cub I took about with me all over the mountains for more than four months, and I learned many a thing about bears from him.   Whenever, as he grew older, we were in camp and he wished to get loose, as he would show by pulling at his chain and bawling, I would free him, and then follow to see what he was after.   He invariably made for some bottom land and dug for roots, or nipped off the grass where it grew young and tender.   I have seen him dig more than a foot down into the ground for some root that had not yet sent up its shoot, and—although how the young rascal could tell just where to dig was beyond me—he always found his tidbit.   Moreover, he knew the berry bushes from the others by the same inherited knowledge.   He would reach up and pull down the branches of these to examine them, although there might not be any berries on them at the time, and he never, that I saw, made a single mistake.   He soon came to pay no attention to my presence, and whenever he dug up anything from the earth that I did not recognize, I would open his mouth, take it away from him, and keep it while I examined it.

When the cubs first come from their dens at from two to three months of age, they are about the size of a house cat, and will weigh from ten to fifteen pounds. At from five to six months of age, that is to say, in July or August, they will probably weigh about thirty pounds. Late in the fall, when ready to den up again, they will tip the scales at from fifty to seventy-five pounds. When a grizzly is nearly two years old, that is, in the second fall, when he is ready to den up for the winter, he will ordinarily weigh anywhere from one hundred and fifty to two hundred pounds.

In their wild state the cubs den up with the mother for the winter following their birth, and follow her during the second summer, after which they are cast aside to shift for themselves, and the old lady will again den up alone, bringing out another litter of cubs the following spring. Naturalists, who have studied these bears in captivity, claim that they then breed every year, but I am satisfied that this is not true of them in a state of nature. In the open, one sees as many she grizzlies with yearling cubs as one sees with little spring cubs.

When a litter of young grizzlies come to leave their mother they do not, as a rule, separate, but travel in company for at least a year; and it is, I imagine, this habit of theirs that has given rise to the idea that the full-grown animals are gregarious. I know that in some cases, and I believe in most, if not all, the cubs den up together for their second winter, and it is (or used to be, alas!) a common sight to see them, during their third summer, working together under the undisputed command of the one to whom they had yielded the leadership. I incline to the belief that they do not separate and breed until the following spring.

When not feeding, grizzlies lie up in some dense thicket near a stream, or, if in a region where they are apt to be disturbed, far back in some high cañon. I have seen many of their beds. In the Selkirks, these sleeping places are far above timber line, dug out from the side of steep mountains where there is not a shrub or a bush to screen them, and where they have an unobstructed view of miles of country. Here, as the signs indicate, they return to sleep day after day.

I have sometimes almost thought that these bears, in a way, enjoy the grand view to be had from these heights. Not only have I found their bedrooms high up among the crags and overlooking range upon range of highest mountain, with restful, wide-spreading valleys below; but it has been no unusual experience, while hunting in these high regions, to see an old bear, after feeding for an hour or more far out of reach of my rifle, stroll deliberately out to the edge of some high cliff overlooking all creation, and sit there on his haunches like a dog, swinging his massive head slowly and dignifiedly from side to side. I have already mentioned the grizzly that we called White Jim, on Wilson's Creek in the Selkirks. This old white bear went through this performance nearly every day for three weeks.

Grown grizzlies do not climb trees. And this for a simple reason. They are not built that way. Once in a while I have seen a grizzly cub go up a tree whose branches started at the very ground and grew in such a way as to allow him to step from one to another; but they never climb smooth and straight-bodied trees as do black bears of all ages, and they never encircle the tree with their arms, as such an animal must, in order to climb a smooth trunk. Grown grizzlies will walk out on a leaning tree such as a man might walk out on with rubber-soled shoes, and they are very expert walkers on fallen logs and timber, and often take to them when trailed. But they not only do not, but cannot, climb.

The grizzly is not, as many hunters and sportsmen suppose, a gregarious animal. One may, and, indeed, often does, see several of them feeding at the same time in the same bottom, or among the bushes of a single berry patch. But, except in the case of a she bear with cubs, or of a litter of cubs that have left their mother, but have not yet disbanded, they will always be found to come singly, and to depart in like manner. Indeed, the etiquette that appears to govern these chance meetings is one of the most amusing things about these animals. An old bear will emerge from the bushes surrounding an open glade where several others are already feeding; he will pause and look critically about as though examining the lay of the land and the distribution of the trees and bushes; but he will show in no way that he is conscious of the presence of the other bears; and these, in their turn, will go on about their business, and by not so much as a batted eyelid show any recognition of his arrival. Sometimes, later on, if two of them meet, or clash over some tidbit, their first movement is always one of surprise at the other's presence, and this enforced dropping of their incognito is more likely than not to be followed by the retreat of the smaller of them from the feeding ground.

I have seen as many as nine grizzlies in one berry patch, and as many as five fishing on one riffle of a salmon stream; but they not only came and went singly, but, while there, they gave no outward sign of mutual recognition or even of mutual consciousness.

It is even open to doubt as to whether the males and females travel together during the mating season, and I have never seen full-grown grizzlies living or travelling in company.

The mating time of the grizzly throughout the Northwest extends, according to locality, from about the middle of June until about the first of August. I have been unable to determine whether individual males and females deliberately seek each other out during this time, pair off, and stay together for a month or more, or whether they meet by chance and again separate. I am inclined to believe the latter, as I have never yet seen the two together at any time of the year. That they do not stay together during the winter I am absolutely convinced, and I do not believe that they remain in company for any material part of the summer.

This opinion, like most others expressed in this volume, is based on many observations, no one of which was conclusive, but all of which, taken together, were not to be ignored. I may, for example, cite the following instance: On one occasion, while on a bear-hunting expedition in May and June, I was camped in a part of the Bitter Root range near an old trapper who was trapping bear. It was near the end of the trapping season, and this old fellow was about to take up his traps, as fur was becoming poor. Near where he had one of his traps set we had, on several occasions, seen the tracks of a large grizzly, and he had left this trap to the very last in hopes of catching him.

It was now the mating season, and although there were bear tracks all over the country, we could never find where more than one large bear had gone at a time.

The last morning that he went to look at the trap near where we had seen the large tracks, I went with him, thinking that, as the bears were so much on the move, I might, perhaps, get a shot. There were also some large snow-banks near by, upon which these bears are very fond of lying when the weather is warm, and as I had already shot several bears there, I thought it likely that I might catch the old fellow cooling himself on the snow.

As we came near the pen in which the trap was set we saw the old grizzly rise up just outside of it, but, as we both supposed that he was fast in the trap, we did not shoot. To our surprise, however, after taking a look at us, he bolted for a thicket and disappeared in a twinkling. It was all so sudden and unexpected that we simply stood agape, and as the bear had only to make a couple of jumps to get out of sight, he was safe before we had any chance of shooting him.

But a greater surprise than this awaited us. On coming up to the pen we found a large female grizzly caught in the trap, and chewed, mauled, and pounded to death. From the looks of things there had been a hot old fight. Of course we had seen from a distance that the logs forming the pen had been thrown down, but we had assumed that the bear that ran away had done this before he knew that he was not fast in the trap. The head of the dead bear was chewed to such an extent that her most intimate friends would not have recognized her, and, upon taking off her skin, we found great masses of clotted blood under the hide that showed how fearfully she had been mauled.

Why the male had killed her can, of course, only be surmised; but the facts seemed to point to only one conclusion. The two had come to the pen, as the tracks showed, from different directions. The female had, of course, come first, as she had been caught in the trap, but the male must have arrived at about the same time, or the one in the trap would have dragged it from the pen. Probably, being exasperated by her predicament, the female had been unable or unwilling to reciprocate in love-making, and the male had become enraged and killed her.

This is one of the experiences that lead me to believe that these bears do not travel together during the mating season. It also leads me to doubt the claim sometimes made that they are more pugnacious at this season than at any other; else this one, already enraged and disappointed, would surely have attacked us as we approached him.

I have never yet seen a whole family of grizzlies together; that is, the male, female, and cubs; and I do not believe that they consort together in this way. I am, indeed, inclined to the opinion that the male will kill the young when they are under four or five months of age. I have noticed that a mother bear, when with her young cubs, takes every means to avoid meeting any male bear, and is always cranky and ready to scrap with any other grizzly, be it male or female, that chances to put in an appearance.

On two occasions I have known an old male grizzly to kill and eat a small cub that was tied up with a chain, and once, while hunting bears in the spring, I witnessed the following incident, or, rather, found the evidences of it. A trapper had caught in one of his traps a female grizzly that was accompanied by her two cubs. She had dragged the trap and clog for several hundred yards, where the clog had finally caught in a clump of brush and stopped her. And while thus held fast, an old male grizzly had come along, and had not only killed her, but had killed and eaten the cubs. When we arrived, he was sitting under a tree close by, and we shot him through the head. We found a few scraps of the cubs lying about, and part of one of them was buried near where he had been sitting.

All my observations, as I say, have led me to believe that a free male grizzly will, if he gets a chance, kill his young cubs; but as the matter is not susceptible of proof, I consulted Dr. Hornaday as to what his observations had been on this point, in the matter of bears in captivity, and I give herewith what he says of the matter: "Of course, male bears in captivity would be likely to destroy young cubs during their first six months—if they got the opportunity. It is absolutely necessary to sequestrate the males and give each female a den wholly to herself and her cubs. We had great trouble in keeping our young cubs from getting their paws into adjoining dens and having them bitten off by older bears."

The grizzly does not den up for the winter at as low an altitude as does the black bear, but seeks the higher hills, where he usually goes into his winter quarters some time in November, or perhaps a little later, according to the locality and the weather conditions.

The den is usually in some natural cave, although occasionally it may be made by the bear himself. I have found a number of the winter homes of the grizzly, and he usually selects a dryer and warmer shelter than does the black bear. Black bears will den in almost any place, and while they usually dig a hole under an upturned root, or under a fallen tree, I have seen where they have used natural caves, into which they have scraped a bed of grass or leaves; but the grizzly seldom, if ever, takes his long nap under fallen timber. He usually seeks the higher altitudes along the timber line, and sometimes even goes higher yet. Here, in the cañons among the cliffs, natural caves are found, and into these a grizzly will scrape and drag anything that can be converted into a bed, and, thus hidden and protected, will pass away the several months of winter undisturbed by snow and storm.

Under certain conditions they will dig large holes under big rocks, in which to make their beds. I have examined several such places, and in two instances found where the bears had dug clean through to the other side. I have also seen where, the elements having in the course of time caved in the earth close to the rocks, the bears had pulled old logs and brush over the breaks, and thus repaired the damage. These were evidently favorite places for dens, and the bears, loath to give them up, had done this in order to retain their old bedrooms. My friend, Mr. A. L. A. Himmelwright, of New York, found a den of exactly this sort a few years ago in the mountains west of

the Big Hole country, in Montana. It was rather late in
the fall, with a veneer of snow on the ground, and he found
the den by trailing the bears, an old female and two cubs.

It is sometimes claimed that when the denning time
comes, the male and female go into the same den; but
this, I think, is a false notion. Not only do I believe that
the male bear would kill and eat the cubs after they were
born, but I have never discovered a den that showed evidences of having been occupied by more than the old
female bear. In the case of a barren she bear, a partnership arrangement as to winter quarters may possibly be
made, now and then; but I cling to the belief that no two
full-grown grizzlies go into the same den for the winter.
In the Selkirks, where the bears den along the high mountain tops, and in the spring come down from their winter
homes over the deep snows, it is an easy matter to back-
track them and find where they have wintered. I have
often done this and, having gone into the mountains two
or three weeks before a bear track could be found, have,
when the first track appeared, taken the trail and followed
it back to the den.

I have seen as many as four grizzlies come from one
den. But they were of about the same size, and were
youngsters not over two years old, and presumably all of
one litter that had not yet been broken up.

James Capen Adams, in describing his capture of Ben
Franklin in the den whence the cubs' mother had just
issued to her death, says that, before going into the dark
den to look for the young bears, "I trembled for the
moment at the thought of another old bear in the den; but
on second thought I assured myself of the folly of such an

idea; for an occurrence of this kind would have been against all experience."

I once found six of these dens in a single day in the Selkirks, and not more than one bear had come from any of them, although, while some of the dens were only large enough for one to lie in, others were of a size sufficient to hold several bears had they wished to sleep together. In the smaller caves, because of restricted quarters, I found much hair that had been rubbed off the animal by contact with the sharp rock, as each had changed his position from time to time during his long period of slumber. Now, were it a common practice for bears to den together in the same cave, it seems that they would have done so here; for all six caves were in the same ridge, and the two extreme caves were less than half a mile apart. I thought, at the time, that I would visit these caves in the fall and see if the same den was occupied year after year, but the opportunity to do so never presented itself.

The grizzly is rather a restless fellow just before denning up. The bed is usually prepared beforehand and made ready for occupancy at a moment's notice. After which, long excursions in search of food are often made about the country, some of them to points as far as ten or twenty miles away.

To what age the grizzly lives in the wild state is entirely a matter of conjecture. I am of the opinion, however, that, under favorable circumstances, they live to be from twenty-five to forty years old. I have come to this conclusion from the fact that I once, for twelve successive years, kept track of one identical bear that was full grown when I first met him, and that showed no evidence of old age when I lost sight of him. This was in the Bitter Root range in Idaho. Every fall, for these twelve years, we saw where this old grizzly had made his way up to the main range to den up for the winter; and each succeeding spring we saw his track as he again sought the lower country to spend the summer among the berries and the salmon. That it was the same bear I am certain, for not only was his track a huge one, and not likely to be duplicated, but I have seen the bear himself many times, and on several occasions could have shot him. He was, as I say, full grown the first time I saw him, and there was during all these years no appreciable difference in the size of his track, which measured nearly fifteen inches. Each fall, as we came from these hunting grounds, and made our way out into Montana, we looked for the track of this old monarch, and invariably saw his footprints. Where he kept himself during the summer and early fall I was never able to learn, as I could not, during these times, find even his track.

Dr. Hornaday, in speaking of bears in captivity, says: "The bears of North America generally reach full maturity between the ages of six and seven years. Some are full grown at six years—others not until seven."

I am inclined to believe that in the wild state grizzlies do not usually reach full maturity until somewhat later. I have watched several that inhabited certain localities, and they, I am sure, did not reach their full growth under eight years. Allowing, then, for the time it must have taken this grizzly to attain his growth, and adding to this the time he roamed the hills under my observation, his age when I last saw him would have been beyond the twenty-year limit. Of course, bears in captivity may, on an average, live longer than those exposed to the dangers and vicissitudes of the open. Yet they have not the inducements to live so long, nor do they, I believe, grow so large as those in the wild state. Those in captivity have neither the fields to roam in, nor the streams to plunge into, nor the sunlight that a bear loves, nor the exercise that all bears take so freely in the wilds. All these things work together for the health and vigor of the bears in far places. On the other hand, those confined in man-made dens and pits have the certainty of food, security from enemies, and the vegetating chances of a life of sloth. Nevertheless, something of the allotted age of free bears may be inferred from the known life span of those in captivity. On March 22, 1909, a grizzly was chloroformed in the Central Park Menagerie, in New York, that had been purchased from Barnum's Circus in 1884, and had, during these twenty-five years, been confined in the pits in New York.

That many bears in the open live to what, for a bear, constitutes a ripe old age, I am able to testify. I have seen them so decrepit that they walked like octogenarians, and I have killed those that showed unmistakable evidences of being full of years. Strangely enough, I have never yet seen or found dead a grizzly bear that had died a natural death. In no one of the caves where they hibernate have I ever found a solitary bone, and, although I have more than once seen an aged bear in a certain locality one season, and found the next year that he had disappeared, I have never, even after careful search, found trace of his remains or hint of the manner of his end.

Probably in some hidden cave, in some remote and lonely spot, the old hero of a hundred fierce contests had passed, all unconsciously, perhaps, from his winter slumber into his unwaking sleep.

XX

FOOD AND FEEDING

THE grizzly bear is, in the technical sense of the word, omnivorous; that is to say, he is both herbivorous and carnivorous, fitted by nature and accustomed by instinct to feed upon flesh or to graze upon grass, to dig roots, gather berries, catch fish, and lap up ants, grubs, and other larvæ.

He is, however, taken by and large throughout the range inhabited by him, almost as erratic in his food habits as he is in color. But these frequently astonishing differences in taste and dietary seem to be entirely the result of environment, and, with occasional exceptions, the feeding habits of the grizzly in any given region are identical.

At first sight these divergent traits and habits appear puzzling, but I think that the explanation is, after all, a simple one. The grizzly is not, I believe, much of a traveller, and hence, those of a given region, having for generations been restricted to the obtainable food of their habitat, are found to have quite lost their taste for food natural to the species, but which they and their immediate ancestors have never seen.

Let us, for example, consider the food of the grizzly in

that part of the Bitter Root Mountains known as the
Clearwater country, in Idaho.   First, in the early spring,
he eats the tender shoots of grass that he finds on the hillsides having a southern exposure, and for some little time
this constitutes his sole sustenance.   He then seeks the
streams, where for nearly a month he fishes for salmon.
He is, as we have already seen, a deft fisherman, but at
this time of year he is not as successful as when the later
runs of salmon appear, when the water is lower and
clearer.   However, for some weeks he manages to live on
his catches.   After the salmon run is over, he again be-
takes himself to the hills, where he turns over rocks and
dead trees, tears up stumps, rips open logs, and overhauls
things generally, hunting for ants, grubs, and any larvæ
that he can find to eat.   These suffice him until the berries
are ripe, and after the berry harvest is garnered he turns
to the autumn run of salmon.   The fish now supply his
larder abundantly for another month, and then he once
more seeks the sidehills and forages for any stray grubs or
ants that were fortunate enough to escape his first cam-
paign.   Such is the bill of fare of the grizzly throughout
the Bitter Roots.

Let us now turn to the same animal a few hundred
miles away in the mountains of northern Wyoming.
Here the grizzly does not dig to any great extent for roots,
nor is he the confirmed grass and fish eater that he is in the
Bitter Roots; but, aside from the berries that all grizzlies
love, and the ants and grubs that they never refuse, he is,
spring and fall, very decidedly carnivorous.   In this
Wyoming region there are thousands of head of elk and
other game.   During the winter many perish, and their

bodies lie until spring under the snow. During the fall many are shot, and their carcasses left lying where they fall. These the grizzly feeds on. In the Bitter Root country, strange as it may appear, not one grizzly in fifty would touch a carcass thus found.

Farther north, in the Kootenai region, and throughout the Selkirks in British Columbia, it is again true that not one grizzly in a hundred will touch flesh. In the Kootenai there is little game of any kind, except bears and Rocky Mountain goats. In the higher Selkirks there are not even fish for them to catch, since, on account of the precipitous nature of the country and the number of waterfalls, the creeks cannot be ascended by the fish. In these regions, therefore, the grizzlies live and die vegetarians. They do, however, spend much time in the late fall in travelling along the higher ridges, hunting for Columbian ground squirrels and whistling marmots, and digging these out of their winter homes. These little animals hibernate, as do the grizzlies, but they turn in earlier; and in a region where they are found the bears take great delight in unearthing them, and sometimes will dig out carloads of earth and rocks to secure a small feast of the little fellows. This is the last food the grizzly obtains before he goes into his long winter sleep.

The claws of the grizzly are well adapted for this peculiar sort of work; yet, long and strong as they are, so much of it does he have to do that, by the time he is ready to go into winter quarters, they are worn down to the quick, and not much, if any, longer than those of the black bear. After his long winter's sleep, however, when he emerges from his den in the spring, he is once more armed with strong curved claws, from four to six inches long.   In those parts of the country where he does not have to hustle so hard for roots and ground squirrels—for instance, in some parts of Wyoming, central Idaho, and Montana— the grizzly does not wear his claws down so close.   Yet he seems, on the whole, to attain a larger growth where he has to work most—just as a hen lays most eggs when she is compelled to scratch hardest.

I have already spoken of the old trapper whom we found on the headwaters of Wilson's Creek.   He claimed to be something of an expert, and figured that he was going to reap a rich harvest in grizzly-bear pelts; but in two seasons' trapping he caught only one grown grizzly and one cub, the latter being the one we wounded and chased into one of the traps which he had set in the trail. This man had killed goats and porcupines for bait, had brought in fish, cheese, honey, and molasses; in fact, he had tried everything to be found in the country, and many things that he had packed in from outside; but the bears simply ignored them all, and seemed perfectly contented to eat grass, dig roots, and nip off the tender buds of the soft maple.

In all our experience in this region we found but a single bear that had any appetite for meat, this being the old fellow that ate part of the horse we led in and killed for bait.   On the other hand, it will be remembered that we detached a shoulder from the carcass of this horse, dragged it across the hills for two miles or more, and placed it on a snow slide in full view of camp, where we could watch it with the field-glasses; and that, although this bait remained there until putrefaction set in and the bears kept the grass cropped close in the little opening, not one ever touched it, nor, so far as we could discover, did they even smell of it or look at it.

I should say, unhesitatingly, that the reason these bears do not care for meat is because they have never known the taste of it. In districts where game is plentiful the bears have been accustomed to live for a part of the time on flesh, and will eat it whenever found. But where they have been forced back into the remote fastnesses of the mountains, where meat and fish are scarcely to be had, and where, from the time of their birth, they are plant eaters, it is not surprising that the carnivorous instinct is not at all, or but slightly, developed.

Almost any mountaineer will testify that animals have to be taught to eat food different from that to which they have been accustomed. We are, for instance, used to thinking of the horse as a grain eater, and most people will scoff at the idea of a horse that is afraid of oats. Yet many times, when no grazing was to be had, I have seen horses, that had been caught up wild and broken to packing, pull back and snort with fright at the sight of a feed of grain, and stand all night with oats in front of them and refuse to eat them. So, I think, it is with the bear.

But aside from the grizzlies in one locality eating meat, and those of another refusing it, there are other idiosyncrasies in the taste and food habits of these animals that teach us to be careful how we generalize from local observation. And this lesson may well be taken to heart by both the hunter and the field naturalist. Above all things, it is important that, when we make a note or mention an occurrence, we state the locality. All animals act differently under different conditions, and a knowledge of these is necessary to draw proper conclusions or form just judgments.

All through the Bitter Root Mountains there grow three plants, commonly called the shooting star, the dog-tooth violet, and the spring beauty. These plants may also be found in the Selkirks in British Columbia. In the Bitter Roots the grizzlies are very fond of the leaves of the shooting star, which somewhat resemble the horse-radish. This plant grows on all the marshes, along all the streams, and in any place that is at all wet, and the grizzly feeds on it more or less all through the summer, or until the leaves get too tough. It is particularly sought for by she grizzlies with cubs. But I have never, in the Bitter Roots, known a grizzly bear to dig up the bulb-like roots of either the dog-tooth violet or the spring beauty. In British Columbia, on the contrary, I have yet to find the first leaf nipped by a bear from the shooting star, while I have seen acres of ground torn up by them for the roots of the other plants. Here the argument of opportunity and environment fails, and I can only record the facts without offering any explanation.

The feeding time of the grizzly depends altogether on the locality. In a country where he is hunted and disturbed, he will either leave the region altogether or will come out to feed at night or just at dusk or break of day. In localities where he is free from annoyance, he will feed up to ten o'clock in the morning, and from two o'clock in the afternoon until dark, and perhaps longer.

In the spring, when they first come from their dens,

The Dog-tooth Violet   The Spring Beauty

ROOTS OF WHICH SOME GRIZZLIES ARE FOND

the grizzlies are likely to be found feeding later in the morning and to return earlier in the evening, and at this season I have even seen them out at all hours of the day. But later in the season, after the ravenous appetite of early spring has been somewhat appeased, they will usually, on the one hand, be seen only between four o'clock in the evening and dark, and, on the other, will not remain out long after the break of day.

The grizzly is the only animal I have ever hunted whose habits one cannot depend upon. The fact that you have seen him out and feeding on berries every morning for two weeks, at precisely eight o'clock, is scant reason for expecting to see him to-morrow at the same time. It would be just like him to then come out at high noon—a time when usually no self-respecting grizzly would think of showing himself. In a way I attribute this to the fact that the grizzly is very wary and, among other peculiarities, likes seclusion. He will change his routine instantly if intruded upon, and if he is molested to any extent will leave his regular feeding grounds for others.

The grizzly, as I have said, does not seem to be much of a traveller. He generally, I believe, spends his life in a restricted area of country, and likes to live where he will not have to go far for food. He loves a dark, wooded cañon near good feeding grounds, and, winding across this cañon, his trails will be found. He is also fond of marshes where there is a stream, and where the small willows grow thick and the grass heavy. Near the edge of such a stream he makes his bed, and here he lies up during the hot days of summer; and not very far away will be found the wallows where he has rolled in the mud to escape the flies. But a grizzly, when forced to, will travel far for the food he craves. He will go many miles to feed on berries during their season, if none grow near his own especial haunts; and in the Bitter Roots he makes considerable journeys to reach the salmon streams. The farthest that I have ever actually known an individual grizzly to go is twenty miles, but I am inclined to believe that they will go much farther than this if need be. I have repeatedly seen a particular bear in one locality year after year, and then found him, fifteen or twenty miles away, across a divide or on another plateau, to which, doubtless, he had moved for better feeding. I believe, however, that they invariably return to their old haunts for the winter.

It is perhaps necessary, in this connection, to call attention to the fact that there is as much individuality in bears as in dogs or horses or people. To the unfamiliar observer all Chinamen look alike, and so do all bears. But to me, as to any one long accustomed to go among them and study them, a bear previously seen and watched is as easily remembered and recognized as a man.

In the regions where the grizzly feeds upon the carcasses of game winter killed or left by hunters, he nearly always drags their remains, after his first meal, to some sheltered spot, and buries it, to keep the birds and other beasts from feeding upon it. He is so powerful that he will drag loads that seem incredible over the most difficult ground and for long distances. Over logs and through thick underbrush he drags and lifts his burden until he finds a spot that exactly suits him. Then he digs up the earth, places his booty in the hole, and piles logs and anything movable upon it until he has completely covered it. Where the bears have been little hunted it is a comparatively easy matter to secrete one's self near one of these caches and shoot the animal when he returns to feed. This is usually either late in the evening or in the early morning, say up to two hours after daylight.

I have already described how a grizzly bear dragged a bull elk carcass up an incline so steep that it was next to impossible for a man to climb up or down without hanging to the bushes for support. The elk's body could not have weighed less than five or six hundred pounds, yet the bear seemed to have transported it with ease, and, after placing it behind a large tree and in under the low-hanging branches, he dug out a hole in the side of the hill and, dragging up logs and brush, covered the carcass completely. The reader is also reminded of the game we played with the grizzly on Wilson's Creek, when every night he stole the body of our dead horse, and every day, with infinite labor, and only by bringing elementary mechanics to our aid, we brought it back to its original position.

It is often affirmed that the grizzly carries such a burden by either grasping it in his arms and walking off on his hind legs, or by "throwing it over his shoulder" (Heaven only knows how this operation would be performed), and so making off with it. The notion is of one piece with that that the grizzly embraces his enemies and hugs them to death. As a matter of fact, he carries anything that he can so carry in his jaws, and burdens that are too great for that, he either grasps with his teeth and, turning his head sideways, drags along with him or, turning squarely toward his find, backs away and drags it after him.

When he comes to a log or other obstruction he pulls, pushes, and boosts until he surmounts it.

Grizzlies have jaws like iron. In and about old Indian camps, where the old leg and thigh bones of elk and moose have been left, I have seen these crushed into fragments, and even ground into particles, by the vice-like jaws of these bears; this, of course, for the marrow that was to be found in them. Many think that the grizzly is a habitual hunter and killer of wild game; and in certain localities, and in times past, this may possibly have been true. This we will discuss farther on. I have never, however, in all my experience, found a single wild animal of any kind whatsoever, except the little fellows before mentioned, that I had any reason to think had been killed by a grizzly.

That the grizzly can, and that easily, kill an elk or a moose, there is no sort of doubt. Nor do I deny that such killings have taken place. But I am firmly persuaded that he never attempts it unless it be in cases of emergency or where some exceptional circumstances lead up to it. Should a grizzly happen, for example, to be near a water lick where these animals come to drink, he might, in one of his impatient rushes, strike down one of them, but the animals that might be destroyed in this way are a negligible quantity.

Strangely enough, however, individual grizzlies do, now and again, turn "bad" and take to killing the cattle of the ranchmen. How they acquire the habit it would be hard to say. Probably through some accident, or by a more than usual gift of putting two and two together and arguing from a stolen quarter of beef, to a walking four-quarters of the same. At any rate, it is not a race habit. One grizzly in the neighboring hills will suddenly turn cattle eater, and come by stealth to satisfy his craving, just as one tiger near a jungle village will turn "man-eater" and come by night to seek for victims. When the guilty grizzly is slain the incident is closed.

James Capen Adams gives a most interesting description of ridding a California ranch of one of these depredating bears, and I have myself known of several instances where such losses were only stopped by killing the thief. One of these argues well the tremendous muscular power of these animals. I know one of the men who watched for and killed the bear, and he told me that when, after waiting several nights, the grizzly finally appeared, they let it have its way before shooting, in order to learn its method of attack. It stole up close to a nearly full-grown heifer and then, in a sudden spring, threw one fore paw across her neck, placed the other on her muzzle, and drawing up one hind leg with a single backward shove of its great claws, not only disembowelled her, but tore out all her ribs on that side.

Nevertheless, the grizzly does not, as far as my observations go, hunt for wild game. In certain parts of the country, where there are plenty of elk and deer, he, as I have already mentioned, depends on them to a large extent for his food supply; but—they must first be killed by the hunter or meet death in some other way, such as being winter killed. The grizzly declines to do his own butchering.

Often, in British Columbia, I have sat and watched a grizzly bear and a little porcupine feeding side by side on the grass near the snow banks, neither one paying the slightest attention to the other. I have also, in this same locality, frequently seen Rocky Mountain goats feeding in the same slide with the grizzlies, and have never seen the least evidence of a desire on the bear's part to molest the goats, or of a tendency on the goat's part to fear the bear. Again, years ago in the Clearwater country, in Idaho, I have, with others, sat for hours watching the elk and deer as they came down to the licks; and while there were many grizzlies always to be seen and their tracks were always plentiful, we never found that a single animal had been killed by them. Had they cared, or been accustomed to prey on the elk, it would have been an easy matter for them to hide behind a huge cedar and strike one down. In those days there were hundreds of elk around the licks at a time.

XXI

HIS FIERCENESS

WE are now arrived at a division of our subject where we are to meet what, at first sight, appears to be a tangle of contradictory evidence, and it behooves us to walk slowly, to preserve an open mind, and to keep our eyes carefully attentive to the signs of the trail. On the one hand, we shall find the sincere convictions and repeated statements of early writers, and a century of unquestioning belief on the part of the public. On the other, we shall find the calmer judgments of trained observers, and the overwhelming weight of contemporaneous experience. Were our fathers wrong about the nature of the grizzly? Or has the animal radically changed in a hundred years?

Personally, I believe that we have to answer "Yes" to both questions; but I am convinced that the amount of alteration in the nature of the grizzly is insignificant compared to the extent to which preconceptions of early hunters colored their judgment.

Let me say, to begin with, that twenty-five years of intercourse with these beasts has taught me to regard them with the most profound respect. I would no more provoke one, unarmed, or rashly venture upon any action that my experience has taught me they regard as calling for self-defence, than I would commit suicide. That they will not fight when they think they have to, no sane man would maintain. That, when they do fight, they are not the most formidable and doughty of antagonists, I have never heard hinted. But that they habitually seek trouble when they can avoid it, or that they ever did, I do not believe. Nor, in the authentic records upon which this popular belief is largely founded, and in which it was first put into words, can we find any facts calculated to uphold it.

On the contrary we see, plainly enough, that the white pioneers, even before they had seen a grizzly, were prepared to meet a dragon, and that, when they had peppered a tough old bear or two with their pea-gun ammunition, they found their expectations realized. That the Indians regarded the grizzly as the king of brutes; that the tale of his terribleness had passed into their folk-lore; that "they never hunted them except in parties of six or more"; that they gave greater honor to one of their young men who had killed one unaided, than to him who took the scalp of an enemy,— all this we can well believe and understand. And that the early explorers accepted the Indian verdict and thought it upheld by their own experiences is no less credible. For the grizzly bear, pursued into his fastnesses and attacked with bows and arrows, would be terrible indeed. And hostilely faced by men armed with the muzzle-loading smooth-bores of small calibre and still smaller penetration, he would be an antagonist but slightly less formidable. These things being so, it is scarcely to be wondered at that our predecessors overlooked two salient features of their experiences: first, that they were themselves invariably the attacking party; and second, that, even so, for every bear that stayed to fight them, there were one or more that ran away.

Let us pass in rapid review the testimony of Lewis and Clark. On April 29, 1805, Captain Lewis, with one hunter, met "two white bears"—their first. "Of the strength and ferocity of this animal," the journal proceeds, "the Indians had given us terrible accounts." (They had told them, among other things, that "they rather attack than avoid a man.") Yet "hitherto," says the journal, "those bears we had seen did not appear anxious to encounter us," and, when Captain Lewis and the hunter fired at these two and wounded both, "one of them made his escape; the other turned upon Captain Lewis and pursued him seventy or eighty yards."

The next record is that of May 5th, when "Captain Clark and one of the hunters met the largest brown bear we have seen. As they fired it did not attempt to attack, but fled with the most tremendous roar." On May 11th Bratton (the man who had boils), wounded his bear and was pursued by it for half a mile. On May 14th the rear guard of the party saw a large grizzly asleep some three hundred yards from the bank of the river, and six of them crept up within forty yards or more and all fired at once. The furious animal charged them and was killed with difficulty. On June 12th they killed two large grizzlies, each at the first shot. On June 14th Captain Lewis had his experience with the grizzly that came upon him when his gun was not loaded, and, after pursuing him so long as he ran away, fled itself as soon as he faced round upon it. This incident (so puzzling to the writer of the journal and so interesting to us with our more intimate knowledge of the species) we will discuss later on. On June 27th we find the account of the hunters who climbed a large tree and raised a shout, whereupon a grizzly "rushed toward them" and was killed. Also the account of a grizzly having come within thirty yards of camp in the night and stolen some buffalo meat. On June 28th "the white bears have become exceedingly troublesome." They "infest the camp at night." "They have not attacked us, as our dog, which patrols all night, gives us notice." Yet the party is "obliged to sleep with our arms by our sides for fear of accident." Then, further, we have instances where men, carrying meat to camp, have been either frightened or confronted by bears, but no actual mention of attack. And the further accounts deal with the grizzly in the mountains and on the Pacific slope of the Rockies, and speak of them as much less fierce than those first encountered.

Now there is little in these recorded facts that differs materially from what I should expect to find among grizzlies, attacked under similar circumstances with similar weapons to-day. It is true that I should expect even fewer of them to show fight, and it is in this regard that I should feel inclined to answer "Yes" to the question as to whether the nature of the grizzly has changed in the past century. I have seen elaborate arguments upholding the theory that contact with man has changed this bear from a savage and aggressive brute to a wary and cautious animal, but my own opinion is that contact with man has merely added to his native caution.

I have met the grizzly under many circumstances and in many places. I have hunted him where he had been but little disturbed, and had seldom come into contact with man, and I have seen him change his habits as his range was encroached upon and his existence threatened. But I have never found him the ferocious and ill-natured brute that he has the reputation of being. On the contrary, I have always found him wary and alert, ready to give one the slip if possible and able to tax one's ingenuity in matching his cunning. He adopts new wiles as new necessities are forced upon him and becomes more cautious as greater caution is required; but even in the wildest and most untrodden portions of his range he is no more on the lookout for a scrap than any other wild animal. I think the fact that now and then an ugly, pugnacious brute is encountered is merely the exception which, if it does not prove, certainly does not invalidate the rule.

I am very far from wishing to assert that the grizzly will not fight. That would not only be untrue but would be a most dangerous assumption to act upon when dealing with him. When it becomes necessary, or when he thinks it is necessary, there is no animal of his size that can put up a fight to equal him. Nothing but instant death on his part or, occasionally, a quick, powerful and effective counter attack, will arrest one of his mad charges. When brought to bay by dogs it is very dangerous to go near him, as he will then charge everything that moves and every bush that shakes. And a she grizzly with cubs is at all times an uncertain customer.

But there are not half a dozen instances on record

where men have been attacked by the grizzly bear in which the offender, if arraigned before a jury of his peers, could not have successfully maintained a plea of self-defence.   Of course, in judging the bears, one must take into consideration the view-point of the bear.   A mother with cubs who charges an intruder approaching too close to her; a sleeping bear over whom a man stumbles in a wood and who strikes him down,—these must be given the benefit of their own doubts.

I do not think that this aspect of the grizzly's nature can be better summed up than by Dr. Hornaday's excellent dictum: "The grizzly's temper is defensive, not aggressive; and, unless the animal is cornered, *or thinks he is cornered*, he always flees from man."

In short, the notion that the grizzly roams about seeking for whomsoever he may devour, is pure nonsense, and that, ordinarily, he will attack on sight, I believe to be equally a myth.   Nevertheless, as far as we can judge, those animals that lived a century ago along the banks of the Missouri and Yellowstone Rivers in Montana, seem to have been, as Lewis and Clark declare them, more courageous and less wary than others of their species found elsewhere, either then or since.

These writers ascribe this to their inhabiting a game country and having become used to slaughter.   But we are wholly unable to judge how far their evident belief that these bears preyed habitually on the buffalo and other game is of a piece with their other misconceptions of them.   They met them in the neighborhood of buffalo, just as I have met them in the neighborhood of elk; and that they were flesh eaters is proved by their stealing meat as, many a time, their descendants have proved the same fact to my inconvenience. But beyond this the evidence is pure hearsay. However, six hunters having approached unnoticed to within forty paces of a sleeping grizzly is pretty good evidence of that animal's sense of security, and another grizzly's having rushed out to investigate the shouts of men in a tree proves the same thing, if nothing more.

Whether they killed buffalo or not, however, they were believed guilty of other acts of which we know them to have been innocent. We can see from several entries in the journal, the wholly natural but wholly mistaken trend of the writer's convictions. The bears, they tell us, "infested their camp at night"—and many a time have I been able to say the same—but the writer evidently regarded the camp's immunity from "attack" as due to the watchfulness of the camp dog; whereas, if the bears had really been bent on slaughter, and had not been merely the prowling thieves and curious investigators they were, the dog would have been but an appetizer for their feast.

On one of my first hunts for grizzlies we were camped beside a little stream which ran down the mountain to quite a large tract of flat land in a kind of a cañon. This creek was lined along both its banks with the black haw, wild cherry, and Sarvis berry, and we had hunted through these thickets in the hope of getting a shot at the bears, whose numbers could be readily inferred from the way the bushes were bent, broken, and twisted, and from the great number of tracks to be seen along the creek. After spending two days at this kind of hunting and not having any success, we decided to move up near the head of the creek in the mountains, as we knew that the bears lived up there and came down at night to feed on the fruit. We accordingly took our camp outfit, which consisted of a frying-pan and our blankets, together with a little "grub" and the saddle horses, and, going as far up the mountains as we could with the horses, bivouacked near the headwaters of the stream. We hunted the mountain all day and, while we did not get a shot, we saw eleven bears, and at night we returned to where we had left the horses, built a small fire close to the creek, ate such food as we had brought along, and then rolled ourselves in our blankets and were soon asleep.

Some time in the night I was awakened by the plunging and snorting of the horses, and, as I listened, I could hear the breaking of small twigs beneath the low bank of the creek, which was not more than twenty feet away from where we were lying. Then I heard the horses break their picket ropes and lope away into the night, and I lay there, wondering if any of the other fellows were awake. There were four of us in the party, all sleeping against a fallen tree, but no two of us under the same blanket. The first sound passed away down the stream, and soon, far up the creek, I heard more snapping of limbs, then an interval of silence, then the scuff-scuff of some heavy animal as it came along down the trail. And the shuffling feet among the pine needles came nearer and nearer until it passed not more than ten feet from me. I drew a long breath of relief and started listening again, and soon I heard another. And there I lay while four bears passed, and as I was on my first hunt for grizzlies, and, of course, thought of all the bear stories I had ever heard or read, I fully expected, as each bear passed me by, that the next would be looking for just such an innocent, unsuspecting idiot as I felt myself to be. But, strange as it appeared to us next morning, none of us had been molested; and, although one of the party declared that a bear had stepped over the log near his head (he had shrunk into his blanket awaiting in terror the attack that never came), and the huge tracks were all about us, the bears had simply, in passing down the creek to their feeding grounds, run into and through our camp, which we had foolishly pitched right in their path.

But the bears did not return that way. And, although we camped in this same place for two more nights, they never used this trail again while we were there. The horses, which were evidently alarmed by the smell of the bears as they passed along, had broken their ropes, fled out of harm's way, and then stopped and gone to grazing. We found them only a short distance away and the bears had paid no attention to them.

Since then, in sleeping in the open near their feeding grounds, I have often heard grizzlies in the night; and have often had them "infest" more formal camps, coming to steal and hunt for meat or other food. Indeed, as already told, I have had hard work at times to protect my supplies from them. But although it has been mighty seldom that I had a dog around, I have always enjoyed the same immunity from attack as did my earliest predecessors in grizzly-bear study.

It is also quite evident from the journal of Lewis and Clark that they unquestionably assumed, whenever a grizzly approached or followed a man, that he did so with sinister purpose, and they "dared not send one man alone to any distance, particularly if he had to pass through brushwood." But we know now that the grizzly is chock-full of curiosity, and that one of his habits is to follow up any trail that puzzles or interests him, be it of man or beast. This trait has been noted and misconstrued by many of the early commentators, and even Adams, in speaking of the Rocky Mountain grizzly, says: "He is more disposed to attack man than the same species in other regions, and has often been known to follow on the human track for several hours."

So often have I seen this curiosity and proved it to be innocent, that I have no fear whatever of these animals when indulging this weakness of theirs. Time and again I have allowed one to approach within a few yards of me, and no calm observer who had watched a bear defying his own caution to satisfy his own inquisitiveness, could mistake the nature of his approach. But a man, filled with a belief in the grizzly as a ravening and savage monster, could, on the other hand, hardly fail to mistake his intentions. The accompanying photograph shows a young bear that had got scent of my camera, and had turned out of his way to see what the scent might mean. One can see his companion in the background following the trail. That he looks savage I think you will agree. In short, you would "hate to meet *that* on a dark night." And yet, as a matter of fact, half scared and wholly curious, he was ready to bolt at a moment's notice, and did bolt the moment the flash terminated his stalk.

Under such circumstances as these my own rule of

From a photograph, copyright, 1909, by J. B. Kerfoot

A CASE WHERE CURIOSITY LOOKS LIKE BLOOD-THIRSTINESS

conduct is a simple one. If a grizzly approaches me knowing that I am there—or that *something* is there that he wants to identify—I have no fear of him; but I have a very wholesome fear of allowing one to approach me unwittingly, so that a sudden discovery of me might startle him or appear to him like an attack. If one retreats before a curious grizzly the chances are that he will follow, and then, to a misinformed observer, his intention of foul play seems proved beyond a reasonable doubt.

These facts being understood, it is most interesting to recall the experiences of Captain Meriwether Lewis on June 14, 1805. Certainly a braver man than Captain Lewis never faced a bear, and this fact (knowing what we know) adds an element of humor to the scene. On that day, it will be recalled, Captain Lewis had shot a young buffalo, and, without having recharged his muzzle-loader, was waiting for the animal to bleed to death.

Suddenly "he beheld a large brown bear which was stealing on him unperceived and was already within twenty paces." Remember, please, what the captain had been told about this animal; remember how long it took to reload the gun he carried; look again at the photograph of a curious grizzly "stealing up unperceived," and imagine his state of mind! He was on an "open, level plain—not a bush or a tree within three hundred yards." The river bank was low and offered no concealment. There was hope only in flight, and not much hope in that. In this dilemma the captain "thought of retreating at a quick walk as fast as the bear advanced, but as soon as he turned the bear ran open mouthed and at full speed upon him."

Now Captain Lewis was a brave man and likewise a truthful one. We do not, for an instant, question his complete sincerity. But, in view of the sequel, we may, perhaps, query the "open mouthed" and question the "full speed." For, starting with a lead of twenty paces, Captain Lewis was still twenty feet ahead at the end of the race, and an "open-mouthed bear at full speed" would have had him in six jumps. Be that as it may, however, the captain, when he had run eighty yards, finding that the bear was gaining, bethought him that he might stand a better show if the bear had to attack him swimming.

"He therefore turned short, plunged into the water, and facing about, presented the point of his espontoon. The bear arrived at the water's edge within twenty feet of him, but as soon as he put himself in this posture of defence the bear seemed frightened and, wheeling about, retreated with as much precipitation as he had pursued. Captain Lewis returned to the shore and observed him run with great speed, sometimes looking back as if he expected to be pursued till he reached the woods."

Now it may be thought that even a ferocious bear might be terrified by the "presented point of an espontoon"; but having myself never owned one of these mysterious weapons, and having, nevertheless, seen scores of grizzlies act for all the world as this one did in its retreat, I do not believe that this terrible engine of destruction (not mentioned, by the way, in the Standard Dictionary) had anything to do with it. Indeed, I have no manner of doubt that if the captain had thought of his espontoon in the first instance, or if he had so much as waved a hand above his head, the encounter would have ended then and there.

The description given by Drummond, the botanist, of his experiences with grizzlies in the Rockies in 1826 coincides exactly with my own observations of these animals fifty years later. He noted their curiosity—often came upon them standing up to look at him—but found that if he made a noise with his specimen-box "or even waved his hand" they ran away.

One glimpse, too, we get of the actual descendants of the Missouri River bottom grizzlies of Lewis and Clark; and this, as far as it goes, tends to suggest that fifty years had left them much as they were when first encountered. The story is told over the signature "Montana," in the issue of *Forest and Stream*, of December 12, 1903. It seems that on May 21, 1860, a party from the American Fur Company's post at Fort Benton camped on the site of the Lewis and Clark camp of May 14, 1805. This party included Malcolm Clark, trader, a big giant of a fellow; John Newbert, tailor, and one Carson, cordelier; and Clark, having followed a grizzly into the woods, came upon him standing up to look back at him, and shot him high up through the lungs. Clark was armed with a Hawkins muzzle-loading rifle, shooting a ball thirty to the pound. The wounded bear charged instantly, knocked the clubbed rifle from Clark's hands, and, felling him with a blow of his armored paw, rushed on. Clark, whose skull was fractured, fell in his tracks; but his companions, waiting at the edge of the timber, killed the bear as it came out, and then, bearing Clark to the canoe, carried him two hundred and fifty miles down the river to

Fort Union, where a surgeon trepanned his skull and he recovered.

In 1878 an English sportsman, named Andrew Williamson, visited Colorado in search of big game, and, in an account of his experiences, afterward published, shows himself to have been a keen observer. In his chapter on the grizzly he says: "Though at least one-half of the stories current in America as to the ferocity of the grizzly bear do not deserve credit, yet . . . he must be regarded as by far the most formidable of the wild animals in America, . . . one the sportsman who has not thorough confidence in his aim and his ability to keep cool had better, when encountered, let go in peace. He will, however, if left alone, unless suddenly stumbled on at close quarters (when he will get on his hind legs with a 'Hough! hough!' calculated to try the strongest nerves) or if it be a female with cubs, rarely if ever attack man, but will, on the contrary, beat a retreat with all the haste he can."

Mr. Williamson's method of hunting, developed through his observing that the grizzly in that region "seemed to prefer, when he could get it, the carcass of any dead animal, no matter how putrid," was to put a bait near a trail and watch from a place of concealment. But the grizzlies were altogether too cunning to fall into this sort of ambush, and he repeatedly mentions finding, by the tracks in the snow, where they had waited for his departure, followed on his trail till it left the woods, and then returned to eat his bait.

Mr. Williamson finally took a leaf out of their own book, tracked one of *them* through the snow, and killed it by a single shot between the eyes. Moreover (he was a man after my own heart), he had an 8 × 10 camera with him. and, although this was in the days of wet-plate photography, when he had got his bear, he packed his apparatus to the spot, set up a tent, and built a fire to prepare his emulsion, and took an excellent picture of the dead animal that he reproduces in his book.

The belief seems to be firmly established in the popular mind that the grizzly is markedly more pugnacious and aggressive during the mating season than at other times. However, not only has the personal experience of many years led me to believe this idea to be unfounded, but I know of only a single recorded instance that can, even remotely, be said to uphold the claim.

This is the experience of Drummond the botanist, who, in the latter part of June, 1826, in the Rockies, saw a male grizzly caressing a female. Soon after he noticed that they came toward him, but whether by accident or to attack he did not wait to see, but climbed a tree. He then shot the female, "whereupon the enraged male rushed up to his tree and reared against it, but did not try to climb." The bear then returned to his fallen mate and Drummond shot him also.

Now here, as usual, the witness was the attacking party, and the bear's pugnacity only developed after the overt act; so that, although it is possible that the same grizzly, if disturbed and fired upon while feeding instead of love-making, would have run away, the evidence is somewhat slight to uphold a sweeping statement in regard to the species.

On the other hand, we have James Capen Adams's

description, already quoted in full, of the peaceful and repeated visits to his camp by Lady Washington's lover. This also is an isolated incident and no general deductions should be made from it.   But as the two occurrences stand face to face and, so far as I am aware, make up the whole of the direct testimony in the case, we are scarcely warranted in finding a verdict against the grizzly.

As for myself, I have been in the wilds during the mating season of these animals in many different years, and in many different parts of their range, and I have not only invariably found them as hard to locate and as difficult to approach at this time as at any other, but in the few instances when I have seen anything bearing on the point under discussion, the evidence has borne against the assumption of any added fierceness on their part.

Adams speaks of the grizzly as terrible if wounded or cornered, and, using as he did the ineffectual weapons of the fifties, he had many hand-to-hand encounters with them.   Yet he tells of many that ran away, and of one that even ran away from her cubs to get out of the way of danger.   I have myself known two that ran away and left cubs to their fate, but this is the rare exception, and most of them will fight for their young to the last spark of life.   As for the others, I have always found about as much individuality among grizzlies as among people. One, if wounded, will fight to the last gasp, and the next will try with its last breath to crawl away.

But I have never known of a single instance where one of these bears turned out of his way, unprovoked, to attack a human being.   I have known several cases where men have been wounded, and one instance where a man was killed by a Rocky Mountain grizzly, but no one of these cases would I call an unprovoked attack.

The man who was killed was out prospecting and carried no weapons whatever. He had a miner's gold pan in his hand, and he and his companion were sauntering along through some down timber. In stepping over a windfall of logs he almost stumbled on an old she grizzly with cubs who was lying beneath the jam, and as he turned to run the old bear rose, dealt him a blow with her powerful paw that smashed his skull, and immediately hustled her cubs out of the logs and made off at a fast pace, paying no attention whatever to the other miner, who was standing a few feet away, bewildered at the misfortune of his companion. I would certainly not regard this as an unprovoked attack, and should expect the same thing to happen to me should I, by accident, commit the same indiscretion. Fortunately the grizzly is so alert that it is not once in a man's lifetime that he would be able to approach thus near to one without the latter's knowing it. And, whether they have young or not, if they are aware of any one's approach, they will scurry away long before the intruder has any chance of seeing them.

In the other case, the injured men were hunters who had followed bears into the thicket after wounding them, and almost any wounded animal will attack one who is foolish enough to trail it into thick brush and down timber. The worst thumping I ever got (it was among my earliest experiences) was from a mule deer that I approached when lying wounded in the brush. I was hurled through the air for twenty feet or more, and then and there learned the wisdom of keeping away from wounded animals.

If, at the present time, the grizzly was prone to attack people at sight, it seems evident that more hunters would see them when seeking them in their chosen territory. There are scores of sportsmen who would give almost any amount to get a shot at one of these "ferocious and awful monarchs of the woods"; and they hunt year after year in countries where bear tracks are more plentiful than those of deer. But the wary brutes remain out of sight and, for the most part, manage to keep a whole skin. I once had a man out with me who said, when speaking of the bear tracks to be seen, that if he were in a country where there was one deer track to the fifty grizzly tracks that he saw, he would guarantee to kill six deer a day.

To sum up, then, it seems to be beyond doubt that a century's contact with men armed with rifles has rendered the grizzly bear a more wary and cautious animal. It would, indeed, be strange if this were not so, for the grizzly is quick to learn and has had innumerable opportunities of learning; and there have been thirty or forty generations during which his individual lessons have been moulding the instinct of the race. But that, during this time, the grizzly has changed from a bloodthirsty and ferocious tyrant to an inoffensive minder of his own business, "defensive, not aggressive," I can find nothing in the records to show, nor do I for a moment believe.

XXII

HIS VITALITY

ANOTHER long-asserted and long-allowed claim made for the grizzly relates to his marvellous vitality. The literature of the subject bristles with statements in regard to his tenacity of life, his ability to disregard awful wounds, and the amount of lead with which he will get away.

The grizzly is now comparatively scarce, the conditions in regard to weapons have greatly altered, and altogether a discussion of the subject is not free from difficulties; but I think that a careful examination of the statements of the most reliable of old-time hunters, a consideration of the conditions, both of mind and of weapons, under which they operated, and a comparison of these with conditions as we find them to-day, will enable us to arrive at a very fair conclusion.

We may, to begin with, throw out of court entirely the writings of the romancers. They naturally seized upon this alleged attribute of the grizzly as one that lent itself admirably to their purposes; and looked upon it as a waste of good material to kill a bear with anything less than a dozen shots and a few knife wounds, and to have the riddled animal pursue the hero for a mile or so, and kill at

least one of his companions before itself giving up the
ghost. The grizzly was their opportunity and they used
him nobly.

But there is abundant testimony remaining, and that
of a more impressive kind. Lewis and Clark hardly
ever mentioned killing one of these animals without dwell-
ing on the ability of the species to take punishment; and
it is made clearly evident that this, as much as any other
fact, contributed to the awe with which they regarded
them. "The wonderful power of life which these animals possess," says the journal, "renders them dreadful, their very track in the mud or sand . . . is alarming,
and we had rather encounter two Indians than meet a
single brown bear."

Now, before discussing the weapons used by these
early hunters and estimating the effect of their very real
dangers upon their judgment, I wish to call attention to
one or two facts, and to one or two inferences that seem
to me to flow from them.

To begin with, I have never seen it claimed that the
grizzly has degenerated in this matter of vitality. Every
writer whose works I have read, while appearing to admit
the accuracy of early observations, takes it for granted
that the perfecting of the modern rifle accounts for any
discrepancies that may appear between those observations
and our own.

Next, I want to note that if the grizzly really had, in
the early days, exceeded all the other animals of his habitat in his resistance to wounds and in his ability to with-
stand the shock of them, this difference between him and
them should have become more marked, not less so, as

these shocks became greater and these wounds more grievous. Yet I have killed well over a hundred grizzlies without finding them any more tenacious of life than many other wild animals. They cannot stand any more punishment than the deer or the elk, and they cannot begin to stand up under the rain of bullets that an old Rocky Mountain goat will survive.

Finally, I would suggest that it is only human nature (especially when badly armed) to be more impressed with the vitality of an animal which, when wounded, takes the offensive, than with the vitality of one that, when similarly wounded, invariably runs away.

Of course, the question of armament is not one to be lost sight of in reviewing the testimony of the early hunters. Their rifles were mostly smooth-bores of small calibre, not larger than the present .32, carrying bullets in many cases seventy to the pound, and all of them were muzzle-loaders with no definite charges of powder. Their penetration, variable under such circumstances, was always slight as compared with the present perfected weapons, and it was impossible for them to drive a ball through the shoulders of a tough old grizzly or even of a young one.

Armed with such a weapon it was necessary to approach very near to one's quarry, the chances of killing a large animal with one shot were small, and it took time to reload. *And the wounded grizzly was a fighter.*

Now it is really not at all distinctive of the grizzly that one attacked with weapons of small range and penetration should, even though "having five balls in the lungs and other wounds," swim half-way across a river and survive twenty minutes; or that one shot once through the lungs should go a mile, lie up in the woods, and be found "still perfectly alive," an hour later. Every hunter of elks or goats could match these instances with others at least equally remarkable. *But these animals run away.*

If one thinks to lose one's life by an ineffectual shot, the refusal of the animal to drop at the first fire is far more impressive than if one only thinks to lose a deer. The difference is psychological, and lies not in the comparative vitalities of the wounded animals, but in the varying effects of this vitality on the man behind the gun.

After giving due weight, however, to these considerations, I am still of the opinion that many old hunters were inclined to draw the long bow when it came to recounting their experiences with grizzlies. Take any old hunters, in either America or the Dark Continent, and some stories they tell beget serious and even amused reflection. In some of Gerard's tales of lion hunting in Africa, for instance, the grizzly bear is put completely out of countenance. In one case this writer tells of a lion hunt in which from two to three hundred persons took part. In half a day's shooting five hundred shots were fired, one man was carried away dead, six were crippled, and the lion was still doing business at the old stand. This either means that the hunters were incredibly poor shots, that their ammunition was worthless, or that the pen is mightier than the rifle.

As already stated, most of my grizzly-bear shooting has been done with a .45 single-shot rifle. I loaded the shells myself and used a hundred grains of powder with swaged lead bullets weighing six hundred grains.

I did not shoot from distances much exceeding fifty yards, and when it was possible to get that close to the game, I could place the slugs just where I wished them, and this was in the centre of the shoulder. The ball rarely passed out on the opposite side, but, unless it was a very large bear, both shoulders were broken and, of course, under such conditions, the animal could not run half a mile and then maul the hunter to death. I could, without doubt, have killed many more grizzlies, but I was averse to wounding game and have it get away and suffer torture, and for this reason I seldom took any chances at long-distance shots. And, again, I took great pride in disproving the theory that a grizzly could not be killed at one shot.

Later, for a time, I took up the .30–30 and found that, while it did good execution, it was hardly the arm for good, clean killing. I have never lost but two wounded grizzlies, and these were both shot with the .30–30. If I were to hunt grizzlies again I would take the .35, which I consider the best rifle for large game at the present time. It has an ideal bullet that mushrooms nicely, and the velocity is great enough to produce a tremendous shock.

Compare such weapons as these with those that Adams used in the early fifties. His rifles were two; one, an old Kentucky arm, carried thirty balls to the pound, which would make the weight of the bullets two hundred and fifty-six grains. The other was a Tennessee rifle that shot bullets sixty to the pound, the bullets weighing one hundred and twenty-eight grains. The amount of powder used we can only guess at as, most likely, Adams did. These were not very heavy weapons, one would think, to kill an animal popularly supposed to pack away anywhere from three to thirty slugs from our modern guns, and then perhaps to climb a tree, wait for the unsuspecting hunter, pounce down on him, and mangle him to death.

Adams writes of crawling up within a few yards of several grizzlies and shooting them and then having them run off.   In one instance he tells of wounding an old female with two cubs, whereupon the mother left the cubs and ran away.   In one or two cases he refers to the vitality of the bears he had shot but he seldom emphasizes it. He does, however, on several occasions, comment upon the general attitude of hunters toward these animals, saying, for instance, of his companion on one trip, "He was a good hunter, but, like most of them, not over fond of a grizzly bear"; and of another (named Wright, I regret to say), "He was a good-enough hunter of deer, but, like all other men who have had little experience with them, terribly afraid of a grizzly bear."   He does, however, mention one bear as having been shot through the head, the heart, and the bowels, while several balls had taken effect in the sides, but had not gone through the fat. The bear ran seven or eight hundred yards after this much shooting.   The ball through the head could not, of course, have hit the brain, and the one through the bowels would not necessarily have stopped her under the distance mentioned, while those sticking in the sides, passing through the hide only, counted for nothing.   The only shot that could really have proved effective was the one in the heart, and as Adams does not state that he examined this organ to find just where the ball lodged,

From a photograph, copyright, 1909, by J. B. Kerfoot

IN THE PINK OF CONDITION: YOUNG, SLEEK, VIGOROUS, AND ALERT

or what part of it was penetrated, we must assume that it was not the upper part of it, where all the large blood-vessels are. These could not have been severed, or the bear could not have run so far.

This brings up another important and very interesting question, namely, how far can an animal travel after having been actually shot through the heart? The answer is, that it depends altogether upon which part of the heart the ball hits. Some time since, while talking with Dr. Alexander Lambert of New York, who has had a wide experience in hunting large game, I asked him if he had ever known an animal to travel any distance after it had received a shot through the top of the heart, severing all the large blood-vessels. He replied that I was the first hunter he had ever known that made any distinction in heart shots, and that he had never seen an animal survive such a shot as I described more than momentarily. Once he so shot a caribou that was standing still when the shot was fired, and the animal made a couple of spasmodic leaps and covered some fifty feet after being hit. But even this was exceptional. As a rule, an animal so hit drops in its tracks.

On the other hand, the doctor brought out a work on the surgery of the heart and lungs, and showed me that shots in the heart other than the one mentioned were not even necessarily fatal. The following are some notes taken from this book:

"Paré (1552) tells of a duellist who ran two hundred paces after receiving a sword thrust through the heart making a hole large enough to admit the finger, and who fought in a vicious manner all the way.

"Balch (1771) had a patient with a rifle ball in the heart who fully recovered in six weeks and lived eighteen years.

"Hally (1878) had a case of a rifle ball in the heart where the man lived fifty-five days; then death was caused by working in the field.

"Dudley (1882) had a patient with a pistol ball in the heart who lived four days.

"Ferris (1882) reports a case of a man living twenty days with a skewer completely traversing the heart."

So we might go on and show more cases where heart shots have failed to kill at once. But a shot through the upper portion of the heart, severing all the large blood-vessels, produces instant collapse in most cases, and death within a few seconds always. The doctor explained these facts by saying that if an animal collapses at once from a shot in the ventricle or lower portion of the heart it is from the shock. Otherwise death only results from the comparatively gradual filling of the pericardial cavity, and the consequent smothering of the heart's beating in its own blood. The shattering of the auricle or upper portion of the heart, on the other hand, not only severs the large blood-vessels and almost instantly floods the pericardium, but paralyzes the heart's action at its source, since the impulse of this action starts in the auricle.

Of course no statement made in regard to a wound has any scientific value unless the wound was carefully dissected out after death. And no statement has any value whatever when it relates to wounds inflicted on an animal that escaped. I have already drawn attention to the liability to mistake inherent in this kind of a report in the case of Jack and the charging grizzly.

The truth is, that the grizzly is just as easily killed with the modern rifle as is the duck or snipe with the modern shot-gun. The main thing in the whole business is to keep cool and put the bullets in the right place. If the hunter will land well in the centre of the shoulder or just back of it or at the butt of the ear, he will kill at once in every case.

XXIII

FACT *VERSUS* FICTION

WHEN I first began actually to hunt the grizzly I found that much of what I had read about him and most of what I had heard was fiction.

From childhood I had read every book that I could lay hands on that treated of these bears, and later I had listened (I dare say with open mouth and eyes) to those I met who claimed to have had experience. I had come to look upon the old-time hunters as heroes and demigods, and was inclined to accept their successors, when I ran across them, as teachers at whose feet I was glad to sit and learn. When, therefore, my early experience began to tumble my supposed knowledge about my ears, I hastily said in my heart, like the Psalmist, that all men were liars.

But since then I have seen more both of men and of bears, and have come to realize that if the men who have written nonsense about grizzlies were technically liars, most of them were quite unconscious of the fact; and that if grizzlies are not altogether as they have been represented, they are sufficiently variable and individual in their actions and habits to have, in most cases, supplied some nucleus of fact for the fictions to form on.

I have come, for instance, to see how inevitable it was that, with the exception of here and there a really scientific naturalist, hardly any of those who have written about the grizzly have written from personal experience. And I have come to understand how naturally, under these circumstances, more romance than truth has found its way into print, and why it is that so very little of what is set down actually touches the real character of the animal. And I have thought in this chapter to speak of a few of the more widely current of these misconceptions, and to cite a few amusing instances of their method of growth.

And first let us quite candidly face the simple truth that, as a rule, the old hunters and trappers, however well meaning they may be, are not to be relied upon for information that is worth much from a scientific standpoint. I well remember the first one I ever saw. He was an old, grizzled fellow, all covered with scars, which he claimed were the results of his encounters with grizzly bears, mountain lions, and Indian arrows. This old chap had heard that there was a man in town that was going bear hunting, and he took occasion to seek me out and have a talk with me about the trip.

He said that as sure as I went hunting grizzlies with the gun I then had (it was my old .44 Winchester) I would be killed, as it was not powerful enough to kill a bear. He declared emphatically that no bear could be killed with one shot, and that the animals would attack a person at sight. He maintained that he had shot grizzlies that had gone a mile or more after receiving several mortal wounds, and that, when finally overtaken, they were found to have plugged up the bullet holes with moss to stop the flow of blood.

When I returned I hunted the old fellow up, and told him that the bears were too wild to hunt with any show of success; but he merely looked me up and down, remarking that this was my first hunt, and intimated that if I kept on hunting and remained of the same opinion, people would not be bothered long with my presence above ground.

And I dare say that up to a certain point he was honest with me. These old fellows are as full of superstition as an egg is of meat. There are a hundred bits of wood-lore and animal legend that they have taken on faith, and that, not being at all vital to the conduct of their own affairs, they have never even questioned and would never think to question. They are quite devoid of what might be called scientific curiosity. The one thing about a bear that interests them is his hide. The only facts they ever learn about him are how to lure him into traps. If this old man had ever really shot a grizzly, had ever come into closer quarters with one than to set a deadfall for him, I have no doubt that he looked back upon the adventure much as St. George may have looked upon his set-to with the dragon; and the tale of his prowess had grown in the telling until he believed in the revised version himself. I have heard many of these old fellows declare that the mountain lion of California has a mane like the African lion, and that they had killed these animals that would measure from twelve to thirteen feet from tip to tip.

And, of course, we must not mix up the entirely distinct acts of lying and "stuffing the tenderfoot." When a man can neither read nor write, and lives most of his life alone on fresh venison and flapjacks, he is entitled to some amusement.

One of the most widely disseminated legends about the grizzly is the alleged fact that they bite and scratch trees as a sort of challenge to would-be rivals. It has even been asserted that these marks are purposely made and duly heeded as a sort of "warning against trespassers," and mark the limits of the range claimed by the bear that makes them. The laws that govern the matter have almost been codified. We are told that the grizzly that posts one of these notices holds a good title to the posted territory until another bear comes along that can put his own mark above it. That the bear with the longest reach is "boss" of that ward. We are told how an ambitious young grizzly on the lookout for a location will wander from one part of the hills to another, measuring up the various marks, until he finds one that he can overtop by an inch or so, when he puts his sign-manual above it and enters then and there into possession while the old owner slinks off to look for a new job.

One writer has even told of an especially clever but dishonest young bear that rolled a stump up to the notice tree, and by standing on it placed his mark so far out of reach of ordinary property owners that it struck terror to a whole neighborhood.

Now there is just one grain of truth in this entire mass of imagination—grizzlies do, occasionally, bite chunks out of trees. Why they do it the Lord that made them may know, but I am certain that no one else does; and, so far from ever having seen it claimed that the writers of these phantasies ever saw a grizzly examine one of these "challenges" and heed it, I have never seen an eye-witness's account of the making of these marks; and only three times in all the years that I have watched these animals have I stumbled on a sight of the operation.

The first time was many years ago while sitting one evening on the side of a mountain watching five grizzlies that were out feeding. It was August, and their hides were worthless, and I was studying them with no idea of shooting. The bears all came out of a thickly timbered cañon through which ran an old game trail made by deer and elk. I had often travelled this trail, and noticed that here and there large pieces had been ripped out of the trees, apparently by some animal with long, sharp teeth. Inspection indicated that two teeth had been sunk into the wood an inch or more, and then, by a sharp twist, a slab of wood had been torn off, and I had supposed the animal that did it was a bear, for I could plainly see the mark of a bear's claws in the bark.

The evening in question, while watching the mouth of this cañon, I saw, first, an old she bear and her two yearling cubs appear and start feeding on the ripe berries. Later an old male came out and also started feasting; and these four bears were making great inroads on the berries when a fifth appeared. He was smaller than the older male, but he came out of the cañon slowly and sedately, and seemed to be very lazy and not more than half awake. He came to the edge of the timber and, looking indifferently around, as these animals will under such circumstances, sat down and scratched his ear with his hind foot. He then got up lazily, sniffed up and down the trunk of a small fir-tree, stretched his paws upward and, raising himself on his hind feet to his extreme height, set his teeth into the small trunk and yanked off a chunk similar to those I had seen scattered along the trail. This was all done in the most unconcerned and bored manner imaginable, without any show of ugliness or temper. There was nothing to indicate in the least that the brute intended the act as a defiance or a challenge to any other bear. He acted as if he had nothing to do and was hard pressed to pass away the time. Afterward he walked out to where the other bears were and joined them at berry picking. The other male bear paid no attention whatever to the action.

On another occasion I saw two three-year-old grizzlies peacefully ambling along a side hill. They were tranquilly inclined and were apparently out for a promenade, with nothing of special importance on their minds. They would walk along for a short distance, stop and sniff at stumps, scratch a little, and then move on again. After a time they came to some trees, and one of them stood up with his paws against a trunk, smelled quite around it, turned his head sideways, drove his teeth through an inch or more of wood, and with a twist of his head ripped off a slab. He then sniffed at the open place, lapped it a little with his tongue, dropped down on allfours, and followed the other bear that had meanwhile moved on. In this instance it was the larger of the two bears that did the "challenging."

On the third occasion I saw a lone bear stop beside a trail and go through practically this same performance.

In each instance it was a grizzly that did the biting. I have never seen a black bear make these marks on trees.

I would like, myself, to know why the bears do this, but I never expect to. And, after all, the action is so casual, so animal-like, so similar to a cat's stretching itself against a tree, that it is probably quite without hidden meaning. It seems to me that we are much given to overworking our imaginations.

Another notion commonly current is that a grizzly will throw his fore legs around an antagonist and "hug" him to death. There is no truth whatever in this idea, beyond the fact that a grizzly, in attacking a large animal like a steer, will sometimes hold it with one paw while he strikes it or rips it open with the other. Indeed, I imagine that this supposed habit has been attributed to the grizzly merely because it has long been credited to bears in general.

Again, contrary to the usual belief, I have never yet seen a charging grizzly stand on his hind legs and thus walk up to his antagonist. I do not believe that this is their mode of attack. All that I have seen fight went at things with a rush on all-fours; sometimes with a bawl and a snort and with champing of the jaws, but never with open mouth. They will, however, bite and rend with their teeth, sometimes holding down the object of their wrath with their fore paws while they tear and bite. I have seen them rear up on their hind feet to deliver a blow, but have never known them to do this until they were near enough to strike. The idea that a grizzly deliberately stands up and walks up to his antagonist, like one of the principals in a prize-ring, is a mistaken one.

A grizzly will, upon any pretext whatever, stand up and look about him. Whenever he sees, or thinks he is going to see, anything, up he goes to his full height on his hind feet; but I have never seen one start to make a charge from this position.

Again, it is often supposed that some of the oddly colored bears of the Rocky Mountain region are crosses between the black and the grizzly bears. Any one who has seen the agility with which a black bear will take to a tree if a grizzly happens along, or has marvelled at the refinement of scent or hearing that enables one to detect the approach of a grizzly (and beat a silent retreat in consequence) long minutes before a human watcher becomes aware of a grizzly's presence in the neighborhood, would not need the denial of science to help him discredit this bit of genealogical speculation.

But it would be an endless task to run down all these flourishing misconceptions. Just to give an idea of how they spread, I quote a few extracts from various articles that have, from time to time, been solemnly put forth as authoritative and even scientific.

One writer blandly remarks that "all grizzlies interbreed, and this obliterates some characteristic marks of the several species. On the southern Pacific coast the two gray species—the light and the mud grays—are closely allied." And, again, that "the original silver-tips sprang from grizzly and brown bears, and they combine all the ferocty and prowess of the former, with the agility and stubbornness of the latter."

In summing up the food habits of the animal the same writer says: 'He has a fondness for horse and mule meat, and he will climb trees, rob bird's nests, and eat the eggs there." Also that "he will climb a fruit-tree, strip whole branches of ripe fruit with his huge paws and claws, and then on the way home will finish off the meal with a toad or a lizard." This gentleman also says that "a grizzly loves to feed on ants." But then, perhaps startled over finding himself so well within the bounds of fact, adds that "he knocks the top off an ant hill, buries his nose in the interior, and by a few inward breaths like a suction pump, draws every vestige of life from the greatest hill."

Another writer, a State Senator by the way, tells of shooting a grizzly four times through the heart and having it still chase him over down timber and bad going, and only fall dead as it was about to fell him. And he goes on to tell of a grizzly bear in the San Bernardino Mountains that used to come once a week, climb a live-oak tree, walk out along a horizontal branch over a high-fenced pigpen, drop in, steal a little pig, push the gate open (it opened out), and go home.

One frequently, in the mountains, sees a great fir-tree growing among the rocks with only a thimbleful of earth within reach. If one follows up the published literature on the grizzly bear, one is likely to see that misinformation has the knack of flourishing upon an equally small store of fact.

XXIV

CONCLUSION

I DO not profess to know everything about the grizzly. I do not believe that any one person can, of his own knowledge, know all that is to be known about any animal. Unless, indeed, he has followed and watched and studied that animal in all the different localities and under all the differing conditions where it exists, he is liable to find himself generalizing even in regard to its more obvious habits and characteristics from insufficient data.

Some years ago, while hunting in the Bitter Roots, one of the party who had hunted elk in other parts of the country, and especially in the Olympic Mountains in Washington, asked me if there were any elderberry bushes in the Bitter Roots. I told him that there were many of them. He replied that he would not, in that case, need any one to show him where to hunt elk, as they would be found wherever these bushes grew. So, after much exploration, he selected a large side hill covered with this growth, and there he put in most of his time for a week watching for elk that never came. The other members of the party had killed their game long before this man could be made to believe that the Bitter Root elk were not Olympic elk and did not live on these bushes.

It had always been his custom to cut open the stomachs of the game he killed and to examine their contents in order to see just what food the animal preferred and to know where to hunt it; and he claimed that he had never killed an elk in the Olympics that did not have the leaves of the elder in its stomach. He, therefore, cut open every elk that was killed on the trip to prove that this was characteristic of the species, but never an elder-bush leaf could be found in the stomach of an elk of the Bitter Root range. In fact, these elks seemed to actually avoid the places where this bush grew.

So with the grizzly bear. A naturalist who had studied him in the Bitter Root Mountains would say that he was a skilful fisherman and a greedy eater of his catch, that he fed voraciously on the leaves of the shooting star, and that he seldom touched flesh. And men who had only known these animals in the Selkirks or in Wyoming would, in either case, declare that the author knew nothing of what he was writing about.

But these local discrepancies are the most obvious of our pitfalls. Let the student follow where he will and watch with what devotion he may, the wild beasts guard many of their actions from his eyes. And at the last, when he perhaps thinks that he has surprised all but the most hidden of their secrets, he will come up against that impenetrable barrier that separates their minds from his. Then, if he is wise, he will place a watch on his imagination.

I have tried to the best of my ability to keep these facts in mind. I have avoided where I could the easy mistake of coloring the actions of the observed animal

From a photograph, copyright, 1909, by J. B. Kerfoot

THREE-YEAR-OLDS

with the psychology of the observer. I have put down my experiences with candor, and I have advanced my beliefs without meaning to dogmatize.

One word I would like to say about shooting. I am the last one, although I myself have had my fill of it, to decry the pursuit of the hunter, but if one wishes really to study an animal let him go without a gun: he will learn more about him in one season than he will in a lifetime of hunting to kill. One reason for this is that when one is shooting one will take the first opportunity that is offered to shoot, and this, of course, ends the chances of observation so far as that animal goes. But there is another consideration. It is only when an animal is wholly at ease, unconscious of one's presence, that one sees him as he really is. Then, and then only, do we catch those intimate glimpses and chance views that admit us, as it were, to a knowledge of his home life and to an understanding of the character that underlies his company manners. And my experience has left no doubt in my mind but what there is some kind of telepathy between man and brute as well as between man and man; and that an interested but sympathetic watcher can remain unnoticed where the presence of a hostile one might breed uneasiness, if not suspicion, in the mind of an animal.

Finally, I have dwelt so much upon the difference between the grizzly of popular imagination and the real grizzly of the wilds, that it may possibly appear that my traffic with this magnificent animal has not left me one of his admirers. As a matter of fact nothing could be farther from the truth.

First and last I have hunted and killed all the big

game of this continent south of the Barren Grounds and
Alaska.   Later, as the years passed and I became less
enamoured of killing, I have been interested in the study
of these animals, one and all.   There is, indeed, no form
of life in the open that is not beautiful, and I am not
ashamed to own that I have spent many happy and silent
hours watching the humblest of them.   But not only as a
sportsman did my interest in the grizzly survive the discovery that all my early and romantic ideas about him
were ill-founded, but, as a student, I have steadily added
to my admiration for him.

He is the one wild animal of our wilderness that owns
no natural over-lord.   With the exception of man he
deigns to recognize no enemy.   And if he is not, as he
was once thought, the bloodthirsty and tyrranous auto-
crat of his vast domain, he is none the less its master.
If, in sober truth, he is less terrible than he was painted,
he only loses interest and dignity in the eyes of those
whom fear alone impresses.

In short, just as the grizzly was in the beginning the
lure that drew me to the wilderness, so now, to my mind,
he remains the grandest animal our country knows.

INDEX

INDEX